Gems

A Lively Guide

for the

Casual Collector

Gems

A Lively Guide

for the

Casual Collector

Daniel J. Dennis Jr.

Harry N. Abrams, Inc., Publishers

This book is dedicated to the memory of my parents, Daniel and Rita Dennis, who taught me about love and family values at an early age. Though they are gone now, the lessons I learned as a child remain priceless. I apply these lessons today and every day of my adult life. I hope I can measure up to their example as long as I may live.

Frontispiece: Five green peridots, a yellow heliodor, two blue aquamarines, and a red rhodolite—some of the many varieties of colored gems.

Project Manager: Eric Himmel
Editor: Richard Slovak
Designer: Maria L. Miller

Library of Congress Cataloging–in–Publication Data

Dennis, Daniel J.
    Gems : a lively guide for the casual collector /
Daniel J. Dennis, Jr.
        p.   cm.
    Includes bibliographical references and index.
    ISBN 0–8109–4126–0
    1. Gems—Collectors and collecting.   I. Title.
NK5530.D46   1999   99–11425
553.8'075—dc21

Harry N. Abrams, Inc.
100 Fifth Avenue
New York, N.Y. 10011
www.abramsbooks.com

Contents

A Word of Thanks

If I took the time here to thank each and every person who has helped me get to this point in my life, it would fill another book. Still, there are certain people who simply cannot be overlooked.

First, I would like to thank my beautiful wife, Linda, the kindest and gentlest person I have ever met. She never once displayed even an ounce of doubt about the success of this project and patiently put up with a roomful of paperwork for what must have seemed to her like eternity.

Professionally, I want to extend my appreciation to David Pillow, G.G., for being kind enough to spend a major amount of time reviewing this book for accuracy, and to Carla Mannino, Connie Doerner, and Jonna Myers, who helped guide me through the maze. I would also like to thank the Home Shopping Network, for literally having given me the opportunity of a lifetime. It was from its stage that I was able to develop my craft, and I will be forever grateful for that gift, no matter what the future may hold.

Likewise, I would be remiss if I didn't mention Eric Himmel, senior editor, of Harry N. Abrams, Inc.; my gifted editor, Richard Slovak, for making it all come together; and designer Maria Miller.

Finally, I would like to extend a sincere note of gratitude to all of those fine and caring people who have been part of my nationwide television audience since 1985. To me, you are family, friends, and professional acquaintances, all in one. I wouldn't be here today without all of your support, and I appreciate it more than you can possibly ever know. Godspeed, my friends.

$\mathcal{I}$ntroduction

Welcome to the most innovative work about gemstone jewelry of its kind. Whether you have always wanted to know more about the precious possessions you own or are planning to make your first serious jewelry purchase sometime in the future, this book is just what you have been waiting for. If you have ever experienced the anxiety of buying something you know very little about, then you know all about the "lump in the throat, knot in the stomach" feeling that goes along with the purchase: did I get a good value, or did I get ripped off? This book, with the right combination of facts and folklore, will raise your confidence level and your consumer awareness about gemstones and jewelry to an all-time high. If it were a college course, the book would have to be called "Basic Jewelry 101." But this is not a textbook; rather, it is a user-friendly guide that is easy to read and understand.

I will be the first person to tell you that many reference materials about gemstones have been written by people with technical knowledge and expertise far greater than mine. Although there certainly is a place for books that contain detailed scientific explanations of the refraction of light and the chemical composition of a gemstone, this book instead addresses the needs of the casual gem buyer who does not know a ruby from a sapphire. Nearly every book on gems and jewelry that I have read bombards the consumer with technical language. This one is different, so let the buyer beware: if complex data is what you are after—if your idea of a good book is *Fun with Ferric Oxide*—you might as well not bother going any further, because there is nothing here that even remotely could be construed as scientific or complex.

What this book does provide is a step-by-step guide into the world of gemstones and jewelry. It examines our fascination with them, the many remarkable phenomena produced by their interaction with light, their often surprising sources, ways in which they are enhanced and imitated, and important factors to understand before you buy any. Then comes the glorious parade itself, a stunning

Rainbow moonstones, a relatively new variety of the feldspar group that is famous for its adularescence

assortment of substances—found in rocks, pebbles, and even some animals and plants—that are then cut and polished by experts in a variety of ways so that everybody else can wear them, show them off, or at the very least admire them from afar. The parade is led by magnificent diamonds, followed by the most popular colored gems and then other gems and minerals (some of them pretty rare or even downright weird!), as well as an extended look at the ancient and revered gemstones known as jade. At the end are the exquisite organics, treasures such as pearls, amber, and ivory that come from animals or plants. (Although gemstone jewelry often incorporates precious metals such as gold, silver, and platinum, I decided it would be impractical in this book to give that subject the elaborate treatment it deserves. For that, you'll just have to wait for the sequel!)

In chapter 14, "The Easy Reference Lists," you will find important consumer guides right at your fingertips. Whether you are just looking or in a serious-shopping mode, there are helpful hints and important points of reference designed to rid your shopping experience of uncertainty and anxiety. In addition, would you like to know which gems represent the month when you were born? Did you ever wonder which stones are linked to your astral sign and which are believed to interact best with the planets and other bodies in the heavens? Have you ever thought about looking for gems that can be found right in your home state? Discover these answers and more in "The Easy Reference Lists."

This book is a result of more than ten thousand hours of on-the-air consumer discussions, combined with two years of exhaustive research. I hope that while it enhances your knowledge of gems and jewelry, it will also bring a smile to your face. Remember: "It's always a good day when you learn something new."

Chapter 1

Exploring the World of Gemstones

Our Fascination with Gemstones

From the very beginning, people have had an ongoing love affair with gemstones. Admittedly, in certain eras they took their fascination to extremes. Countries have gone to war, rulers have had their subjects beheaded, loves have been won and lost—all over those tiny rocks, minerals, and crystals we call gems. Archaeological digs support the theory that certain organics, such as amber and pearls, have been with us virtually since the beginning of time. Younger readers may be shocked to learn that some gems were actually here on Earth even before the invention of the video game. Oh, and while I'm at it, one more point of clarification for them: the term *rock* can mean more than very loud music.

In ancient times, gemstones were the sole property of kings and noblemen; commoners were forbidden to own them. In certain societies, if the everyday Joe were caught with certain precious stones, he would be imprisoned or even put to death without trial, for the nobles assumed that the gems had been stolen from people of a higher class. This practice helped give rise to another group that is feared by everyone, even today: lawyers.

Gems have been used to adorn everything from weapons of war and shields to crowns and scepters. Yet—believe it or not—at one time women were prohibited from owning diamonds and other gemstones! After a while, however, it became apparent that the average guy spends about as much time thinking about jewelry as he does about the origin of lint. As a result, once and for all, ladies were put in the position of controlling demand. It almost scares me to think of the condition the jewelry trade would be in today if it had to rely only on the men. Thankfully, the lords of the gem trade came to their senses, and not a minute too soon!

If you are thirsty for more hands-on experience, you can obtain a great deal of knowledge on the subject by visiting a museum. The Natural History Museum in Los Angeles, the Field Museum in Chicago, and the American Museum of Natural History in New York City are just three of the many fine places where you can view impressive displays of jewelry. Few, however, would dispute that the most formidable collection of all is found at the Smithsonian Institution's National Museum of Natural History in Washington, D.C. The Smithsonian has an extensive array of rare and exotic gemstones of historical significance. To see this section of this museum alone would be well worth a visit. If you are planning on seeing all of the different museums of the Smithsonian, allow yourself a few days. If a trip to one of these major institutions is impractical for you, there are many others throughout the United States (and the world) that devote space to gems, rocks, minerals, and similar finds. Even the smallest of displays at your local museum or library can prove to be a pleasurable and worthwhile way to spend a few hours.

New technologies and more advanced mining techniques have enabled just about everyone to have an extensive collection of beautiful colored stones, some of which were not even known to the gem world at the turn of the last century. Even the world of computers, loaded with powerful and detailed software programs, has become an integral part of jewelry layout and design. These new applications now permit not only gems but also precious metals and even costume jewelry to be cut and faceted (given many small flat surfaces, at an angle to one another and often in a particular pattern, to enhance the beauty) in precise, diverse methods unavailable in the past. In addition to its accuracy, this high-tech equipment in some instances offers the consumer substantial cost savings at retail, by replacing the outdated, labor-intensive manual methods of the past. Admittedly, words like *handworked* have a romantic connotation. Still, the substantial savings and pinpoint accuracy combine to more than offset the dreamlike appeal of the handcrafted touch.

Defining a Gem in a Less Traditional Manner

When I first fell in love with gemstones many years ago, an impenetrable boundary separated precious and semiprecious gems. It was almost as if the lords of the gem establishment held diamonds, emeralds, rubies, and sapphires in such high regard that the rest of the gems on Earth were simply allowed to exist. Things have changed, and today the line that once kept the big four apart from the rest of the world has all but completely disappeared. Now we have a new way of thinking, to evaluate gems in a less traditional manner.

Today, gems are evaluated using four basic criteria: beauty, rarity, durability, and desirability. How a gem fares in these four areas determines its value and overall ranking in the gemstone world. Because this departure from tradition

is important not just for gem experts but for you, the consumer, as well, I will take the time to elaborate on these categories just a little.

Beauty

You've no doubt heard the adage that "beauty is in the eye of the beholder." Because it is, this category is the most open of all to discussion. Remember, this is only one of four considerations, so don't get overly taken aback by its meaning. A gem should be pleasing to the naked eye—something that grabs your attention at first glance. The problem here is obvious: what you perceive as beautiful is often different from the perception of your neighbor. All you have to do is spend an hour people-watching and this will become apparent.

Rarity

To determine rarity requires considering not only the mineral group or gem itself, but its quality and its source as well. For example, it would be inaccurate to state that all emeralds are rare, because the gem still exists in large deposits, and stones of less than perfect color and quality can easily be found even today at retail. On the other hand, large, top-quality emeralds of excellent color and proportion are indeed very rare and collectible, commanding thousands or even tens of thousands of dollars. It is also worth noting that some of the rarest gems on the market today come not from the precious-gem category, but rather from those formerly called semiprecious. Think of the natural alexandrite, certain tourmalines, and the highly prized black precious opal, and you will begin to get the picture.

Durability

When considering this criterion, keep in mind two other words that are closely related: *wearability* and *practicality*. If a stone is too soft or too easily chipped or scratched to be set into gemstone jewelry, can it really be classified as a gem? There are many beautiful minerals in nature that evolve into lovely specimens but are not suitable for jewelry purposes, for one reason or another. Such is the case with kyanite, a mineral often found in a deep blue color. Although the stone is lovely to look at, it is easily damaged and nearly impossible to facet, because of its crystal structure. Since it makes virtually no contribution to the gem and jewelry world, does it deserve to be classified alongside the blue sapphire, which contributes so much? Though this, too, is somewhat subjective, it is a point that must be considered when evaluating a gem. As we will soon see, the Mohs' hardness scale is the gauge for durability, accepted worldwide in the gem society.

Desirability

Anyone who has ever looked into a jewelry case already knows that not all gems are created equal. There exists such a staggering span of quality that even within the same type of gem, a stone can be of virtually no value at all, or worth tens or even hundreds of thousands of dollars. This huge gap can sometimes result in one stone being classified as gem grade while its "brother" is branded with the ugly, catchall phrase "industrial quality." This is very much the same as saying your blind date has a nice personality.

You can easily grasp this concept when taking a look at the diamond. Although Australia is actually the world's leader in diamond production, the African nation of Namibia is number one in the production of stones suitable for gem use. In fact, approximately 80 percent of all diamonds produced worldwide are classified as industrial grade! Now, you wouldn't want to pay thousands of dollars for something that closely resembles the lead in a mechanical pencil, would you?

Of course, this fact of classification is not unique to diamonds; it exists in every mineral group found worldwide. As a result, for example, a quarter-carat garnet of great beauty, rarity, and desirability can be worth much more than a heavily included (meaning filled with foreign matter), lifeless emerald of several carats. I think at this juncture it is safe to say that the age-old concept of precious and semiprecious gems has been sent the way of the Nehru jacket. Weep not, for a new day is dawning in the colorful world of natural gemstones.

Assembling Your Personal Gem Collection

If money were no object, this section would not be necessary; so all those billionaires who are currently reading this book have my permission to move on. Gee, how come nobody left?

For those readers who haven't yet attained their first billion dollars, I have separated some of the more popular gems into rough categories of affordability (keeping in mind what I've already noted about the range of quality for each gem). Here, let's assume all the gems we are looking at are good but not at the high end in terms of quality. Since the majority of you will most likely purchase rings more often than any other type of jewelry, let's consider two-carat stones (where applicable) in a simple, four-prong setting of 14-karat gold. (Don't worry, these terms will be explained in chapter 6.) Although there are many more gems than the ones mentioned here, I've tried to identify those that are seen most often at retail. Various factors may affect the strict accuracy of these lists, but they should serve to give you a general guideline as you go forward to assemble your gem collection. Have fun with it, but don't take it as gospel.

Inexpensive ($100 or less): Agate, almandine garnet, amber, amethyst, bloodstone, blue topaz, carnelian, cat's-eye quartz, citrine, coral, malachite, onyx, quartz crystal, rose quartz, smoky quartz, turquoise, white topaz.

Midpriced ($100–$200): Ametrine, blue sapphire, chrysoprase, danburite, fire opal, goshenite, grossular garnets, heliodor, iolite, jadeite jade, kunzite, labradorite, lapis lazuli, moldavite, moonstone, morganite, pearl, peridot, rhodolite, ruby, spinel, (most) tourmalines, white precious opal, zircon.

Expensive ($200–$300): Aquamarine, emerald, (most) fancy sapphires, imperial topaz, indicolite, pink tourmaline, watermelon tourmaline.

Professional athletes only ($300–infinity): Alexandrite, black precious opal, cat's-eye chrysoberyl, color-change sapphire, diamond, mandarin (spessartine) garnet, padparadscha, rubellite, tanzanite, tsavorite.

Chapter 2

Gem Phenomena

Gem phenomena result from the interaction of a gemstone with light. Many types can be observed in the jewelry world. Each and every one is remarkable unto itself, and even phenomenal gems within the same mineral group sometimes differ from one another. One thing that all gems displaying one of these phenomena have in common is their endless popularity with gem collectors worldwide. Let's examine some of the more commonly seen kinds.

Adularescence

Certain types of feldspar exhibit a cloudlike effect that seems almost to float in midair. This is a phenomenon known as adularescence. Consumers are most likely to come across this when searching for adularia and other gemstones in the orthoclase family of feldspars, including the popular moonstone. This effect is caused by the reflection of light in the mineral's crystal structure. Gems showing a white haze, sometimes known as schiller, are more common than those displaying a silver-blue sheen. In either case, gems with a more vivid display are generally more sought after, and costlier, than those with weaker adularescence. Such gems of large carat weight are rare and expensive regardless of color.

It should be noted that you may come across a similar phenomenon in the beautiful blue feldspar known as spectrolite. Although to most of us it may look the same as gems showing adularescence, the correct term in this instance is *labradorescence;* spectrolite is a variety of labradorite, which is from another family (plagioclase) in the feldspar mineral group.

Asterism

Rutiles, or very thin inclusions (foreign particles inside a mineral), that intersect each other at varying angles produce a phenomenon known as asterism, meaning "star." Although asterism occurs in a wide variety of minerals, consumers most commonly encounter those types seen in sapphires, rubies, and garnets. Because of their crystal structure, the star sapphire and star ruby generally have six-pointed stars, while the star of the garnet usually has four points. Other mineral groups such as chrysoberyl, quartz, and tourmaline often can also display an impressive asterism. In both the star and the cat's-eye (see "Chatoyancy," below), proper cutting is critical to the effect.

Aventurescence

Most consumers know little of the phenomenon known as aventurescence. Like the star and the cat's-eye, this effect is caused by inclusions that interfere with light. Unlike the tubular structures that cause the asterism or chatoyancy, the inclusions that bring forth the sparkle of aventurescence are flat. Two of the most popular gems that exhibit this property are aventurine quartz and a variety of feldspar known as sunstone (or aventurine feldspar). The color reflection depends on the inclusion itself; some of the more commonly seen inclusions are copper, mica, hematite, and chromium (an element of the mineral called chromite). Beware of a man-made material usually called goldstone; this material is manufactured glass that is purposely included with small particles of copper.

Chatoyancy

Another phenomenon popular with consumers is the effect known as a cat's-eye. Also called chatoyancy, this is caused by the reflection of light off the naturally occurring, needle-like inclusions called rutiles. As with asterism, these inclusions are usually tubelike in structure, but here they point in only one direction. This property is seen in a wide variety of mineral groups, although some gem purists recognize only the color-change chrysoberyl as a "true" cat's-eye. Still, other gems such as quartz, tourmaline, apatite, and even certain varieties of beryl are known to exhibit chatoyancy. Golden quartz stones that display chatoyancy as a result of inclusions of asbestos are known as tiger's-eye, while those with a blue-gray or blue-green background are called hawk's-eye. A variety of chalcedony known as eye agate was fearfully regarded by ancients as evil—a representation of the common expression known even today as the "evil eye."

Color Change

Gems that actually change color when exposed to different light conditions are perhaps the most spectacular of all phenomenal gemstones. The color-change phenomenon occurs because the shape of certain crystals makes some parts of them absorb light (which is composed of a range of colors) differently from other parts. Most such gems can display two different colors; some even turn into any of three colors (or more), depending on the light. Perhaps the best-known color-change stone is the alexandrite, but it is hardly the only one. In fact, a wonderful color-change sapphire is found in Tanzania, and certain color-change varieties of spinel and even garnet are sometimes seen in nature. Demand for all of these gems is so high that they are well beyond the financial reach of the average consumer, assuming any such stones can be found in the first place.

A different but related phenomenon is called fluorescence, referring to the ability of certain gemstones to change color when exposed specifically to fluorescent lighting. Most often, this phenomenon is associated with fluorite, making it a most interesting gem. In some cases, calcite also exhibits fluorescence.

Note that the color-change phenomenon is not the same as something called pleochroism, which refers to a stone that shows different colors when looked at from different angles in the same light (displaying two colors is called dichroism; three colors, trichroism). Nor does it have anything to do with color zoning, which refers to stones that show two or more colors at all times, such as the bicolored ametrine and the bicolored or tricolored watermelon tourmaline.

Color Play

Many consumers own a stone exhibiting a phenomenon and don't even know it. The color-play properties of the opal, caused by light that is diffracted by (meaning bent around the edges of) tiny particles of silica, are generally accepted by gem experts as still another type of gem phenomenon. The most sought-after color play of all shows red fire against a black backdrop. For additional information on color play, be sure to read the entire section on opals (see chapter 9).

Iridescence

Iridescence, or a display of rainbow-like colors, is often seen in feldspars such as moonstone, labradorite, and spectrolite (but is not the same as adularescence and labradorescence, other phenomena found in some of those stones and described in a previous section), as well as in the variety of chalcedony called the fire agate. This phenomenon is caused by interference with light as it enters the stone and encounters inclusions. Each of these gems gives off its own special glow. To accentuate their iridescent properties, experts almost always cut the gems flat or *en cabochon* (rounded and highly polished, but not faceted). People have both worshiped and feared iridescent gems since ancient times.

Chapter 3

*W*here Gems Come From

"*Y*ou Mean My Pretty Ruby Once Looked Like That?"

If you have ever searched for gems while on an African safari, or have already negotiated gemstone rough (the crude, uncut stone) at gunpoint with representatives of some decidedly unfriendly country, feel free to skip this chapter, for it will hold few surprises for you. There! Now that we have eliminated, say, .0002 percent of our readers, let's talk about the origins of your lovely finished jewelry.

If you have never seen a gem in its natural state, for the most part you haven't missed much. In fact, you could be looking right now at gemstone rough and never even recognize the gem. Before my encounter with a ruby in the rough, I had kind of a different picture of how gems were supposed to look when they are taken out of the ground. You know the old saying "You can't fool Mother Nature"? Well, after examining an African ruby in the rough, I have no doubt that while humans may have succeeded in taking the stone from Mother Nature, she still may have the last laugh. If somebody had asked me, I probably would have given that person the rock for free. My inner instincts tell me I would not have had a prosperous career as a gem miner.

One thing I quickly realized is that this is no place for an amateur; you really need an in-depth knowledge of minerals and how they occur in the wild. After I was exposed to aquamarine rough embedded in a low-grade quartz cavity, and shown a tiny jade deposit almost smack-dab in the center of a yellow rock, I knew that the only way I could ever be a gem miner was if I could take a slide show along to help me identify the gems. But this probably would not be very effective—for one thing, electricity isn't all that abundant in the jungle. Besides, the wildlife would probably eat my screen.

Recovering the Rough Is Not for the Faint of Heart

Extracting these gem findings from nature is a very risky business. It would be nice if you could take a walk through your local park and gather the rough in a neat little wicker basket, kind of like an Easter-egg hunt when you were a kid. Unfortunately, the vast majority of gems found worldwide are located in very hostile environments. Native miners literally risk their lives on a daily basis, venturing out into the wild or tunneling deep into a cavity of the Earth, to bring these treasures to the rest of the world. Wild animals abound and the threat of venomous snakes is often right at hand. Temperatures soar well above one hundred degrees, making dehydration and heatstroke constant companions. Rockslides, cave-ins, even land mines left over from some past encounter with a neighboring country kill untold numbers every year. With all that in mind, is the thought of mining your very own gems still a romantic notion? Personally, I would sooner hunt for seashells on the beautiful shores of Florida.

Primary and Secondary Deposits

When a gem or mineral is found in its original location, this is known as a primary deposit. These gems often are deeply embedded in the ground, so they are usually mined with modern excavating tools. This process is known throughout the gem world as deep-shaft mining. There is evidence that even the earliest miners, before recorded history began, learned to dig craters in order to extract buried gem rough. (Way back when, they also discovered that heating and quickly cooling host rocks could split them apart, sometimes exposing hidden gem materials. Later, they found that splitting geodes, hollow rocks that can resemble potatoes, yields such treasures as peridot and amethyst.) Today, large-scale mining operations use huge earth-moving machinery to extract rough from immense open pits. Of course, such mechanized methods are useless when the deposits are small and scattered.

When gems or minerals separate from the mother rock and end up in an entirely different location, as they frequently do, this is known as a secondary deposit. These deposits are generally formed over time; however, radical changes in the environment, such as volcanic eruptions, can expose secondary deposits quicker than you can say "hotfoot"!

Alluvial Deposits

Gems and minerals that have separated over time from the mother rock often end up in a body of water. This type of secondary deposit is known in the gem world as an alluvial deposit. Because the stones most commonly come to rest in rapids or waterfalls, they are sometimes tumbled to a nice polish by Mother Nature.

As a result, alluvial gems are commonly regarded as higher-grade finds, since much of the weaker, included material has been eroded away. Even back in the Stone Age, the earliest prospectors used crude sifting pans, made of natural materials such as bamboo, to gather such gems from riverbeds and seacoasts. Today, workers literally risk life and limb searching treacherous rapids and waterfalls, often to recover just waterworn pebbles of tiny gem rough. Many of these deposits are found in remote areas of third-world countries, where natives reluctantly embrace this way of life. In reality they have little choice, because many of these out-of-the-way regions offer no other means of employment.

Organic Matter

Besides being found in rocks and waterworn pebbles, some gems are derived from living animals and plants. Gems of this kind are known as organics. Undoubtedly, the most commonly known organic gem is the pearl. This natural treasure forms within many types of mollusks found in both fresh waters and seas worldwide. Although the pearl is the most popular organic gem, there are three others you will eventually encounter: amber, coral, and jet. At one time ivory and even tortoiseshell were considered important organic gems, but thanks to government-imposed limits or outright bans on their harvest, little natural ivory or tortoiseshell is seen in the domestic market today. Because we will take a more in-depth look at organics later on, suffice it to say at this point that each can be found in a wide array of locations and forms.

Tunnel Mining

In order to protect the ecosystems in the delicate rain forests and other parts of Brazil and beyond, that country (like others) has enacted some new get-tough legislation. These laws demand that companies replace any critical vegetation destroyed through the mining process, thus putting needed safeguards in place to protect the land. In order to avoid the high costs of restoration, mining companies now employ a relatively new but perilous mining practice in some places. After opening a relatively small cavity in a mountaintop, natives enter these unstable hallways of nature, risking all for the sake of profit. Tools are carried into the opening by hand, and if no gem rough is found, the men then dig deeper into the mountain from within the opening. This method of excavation is known in the trades as tunnel mining.

The Journey from Rock to Retail

Once the gem rough has been retrieved, whether by deep-shaft mining, alluvial prospecting, or another method, the next step is to separate the gem from the surrounding rock. The force from a huge volume of water usually does the trick, unless there's a very large amount of rock involved; in that case, machines known simply and accurately as rock crushers pound the rock into submission.

Afterward, the gems that emerge are sorted by variety, size, shape, and color, using methods that run the gamut from old-fashioned sorting by hand to high-tech sorting by machines that automatically separate material as it moves along extensive conveyor belts, or by other highly specialized equipment. If you want to impress people with your knowledge of the widely varied sorting process, just drop terms like *electrostatic fields, mineral fractions,* and *rotation chambers*. Your friends will probably pretend to understand what you have just said, even if you don't.

Despite modern machinery, much of the sorting, especially in matching stones by exact shade rather than just similar color, can be an arduous task for even the most experienced gem sorters. Incidentally, some sorters (for example, in Thailand, arguably the gem-cutting capital of the world) are children, who are taught to develop their skills at a very young age with games that involve sorting beans of various colors, sizes, and shapes.

Once sorted, the gems are cut into workable shapes, by everything from big circular saws to hammers and chisels to precise cutting tools for more intricate work. Much of this depends on the size and shape of the material, as well as its composition and the direction of the crystal formations within it. Then each gem is sent to an expert for shaping, faceting, and other treatment to bring forth its beauty. (For specifics on methods of gem enhancement, see chapter 4.) Varieties of faceting constitute the fastest-developing aspect of the gem-cutting craft today, with modern equipment able to place as many as one hundred or more small angled surfaces on a single stone!

To create a mounting for a gem, an artist first draws a design on paper. Powerful high-tech software programs mark out an exact size and shape for the designed mounting, and a mold is then developed for the piece in question. When the item is cast, an experienced craftsperson with an exacting eye for symmetry sets the stone. It is then ready to move—passing, as in every stage, through various intermediaries' hands—to a retailer's display case and ultimately into the possession of a consumer.

Well, that's an abbreviated, nontechnical description of the magical journey that gemstone jewelry takes from rough to retail. Honestly, how many of us ever give any of this a second thought when making a decision to purchase a gem? In fact, you should, because in many instances the type and amount of labor involved account for much of the final cost.

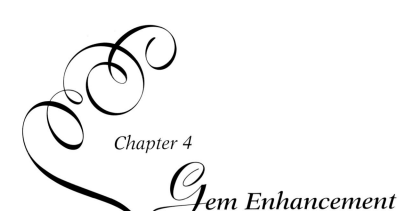

Chapter 4

Gem Enhancement

It may surprise you to learn that nearly every single colored gem on the market today has been color enhanced through one manufacturing process or another. In fact, most of the beautiful colored gems we all know and love so well would not exist without some assistance from humans. The beautiful and popular tanzanite, for example, often comes out of the ground as a murky brown rock that has no market appeal whatsoever; through enhancement, however, experts have been able to show to the naked eye a very pleasing, vibrant blue-to-purple color.

This chapter is designed to educate the everyday consumer about the many processing steps necessary to turn often dull, lifeless minerals into the lovely treasures we all cherish. You will learn a lot, I promise you, without being bored. And please do not view gem enhancements in a negative light. They are as necessary to gemstones as the sun is to a garden of flowers.

Bleaching

Bleaching is the process by which high concentrations of caustic acids are used to treat a gem. Bleaching almost seems to be a paradox: in some instances, it is applied to lighten a gemstone that is too dark, while in other cases bleaching is applied to completely remove the color from a weak, pale gem to bring it to a colorless state. Pearls are often bleached to ensure uniformity of color.

Chemical Treatments and Reactions

Sometimes harsh chemicals and acid baths are introduced to the treatment process under intense heat. This is done primarily to ensure consistency of color. This type of treatment is done selectively and with caution, for only certain mineral groups can withstand such harsh chemicals. If things go wrong during this process, many costly gems may be damaged beyond repair.

The gem we all know under the name black onyx is an excellent example of the chemical-and-heat process. Many consumers are surprised to learn that black onyx is actually not onyx at all but another close cousin, usually gray agate. The agate is soaked in a dense sugar solution, generally for about one month. After the rough has been thoroughly saturated, sulfuric acid is introduced; this creates carbon, which turns the stone black. You can buy your black onyx from the costliest department store in your town, but you are still going to end up with agate. Remember, natural onyx occurs most often in nature as a black-and-white-banded variety of gem, occasionally seen in the jewelry world in ornate carvings such as cameos.

Coatings

The coating process can take a variety of forms. In some instances, an enamel, lacquer, or other liquid coloring agent may be used either to alter the color of a gemstone completely or to complement and unify whatever color is already present. The process can also involve a film, or thin layer of one color or another. In this type of treatment a clear gem, often white topaz, is coated with a micro-thin film, thinner than a single human hair. Applying several different coatings of film can make the gem give off a dazzling display of colors. Incidentally, this variety of coated gem produces its colors through the reflection, rather than the absorption, of light.

Dyes

Gems have been dyed throughout the course of time. Those most frequently subject to this process have included jade, opal, coral, lapis lazuli, pearls, and a wide variety of chalcedonies. Dyeing can be a most effective form of gem enhancement, which can last for many years. Be aware, however, that not all gems take readily to a dye, and poorly dyed stones can lose their color—and their value—almost immediately.

A gem that is dyed should be clearly identified, but some manufacturers and dealers do not. There have been cases in which a lighter-colored, less valuable jade was dyed to a vibrant green color and sold to unsuspecting consumers as the rare and costly imperial jade. So how can you protect yourself? You should exercise common sense and, if possible, deal only with reputable sources whom you (or friends who are more knowledgeable) have relied on in the past. Check the stone for uniformity of color, looking especially for any dark or light spots, which may indicate a poorly dyed gem. And don't be afraid to express your concerns before making a selection. The only problem here is that, as with other treatment processes, the retailer may be too far out of the loop to know for sure what has been done to a gem.

Foiling

Admittedly, your chances of ever encountering a foil-backed gemstone at retail are remote at best; still, it pays to understand this process. Foil-backed stones are usually ones of poor quality and color, set into a closed-back setting that has been lined with colored foil. The setting is always closed so as to leave the foil undetected. This form of treatment is almost always done with a solitaire, or single set gem. The stone is often a lower-grade stone of such weak color that it cannot be classified as a gemstone without some help. In other instances, foiling is used to back a clear gem with a silver foil to create a dazzling, diamond-like look, or with a colored foil to imitate a high-quality colored gem. Although just about any colorless gem can be used in foiling, those seen most often are the white topaz, spinel, danburite, and zircon. There is absolutely nothing wrong with foiling, as long as it is disclosed to the consumer. If you want to be almost 100 percent certain that your gem is not foil backed, stay clear of any sparkling gem in a closed setting.

Glass Filling

Glass filling is a process in which gem rough is heated to extraordinarily high temperatures, up to two hundred degrees Celsius, and then an artificial material (in powdered form) is packed around it. Borax, a cleansing agent used in the manufacture of glass, is the substance of choice, but aluminum and silica are also sometimes used. The intense heat causes the powdered material to liquefy and fill the cracks of the gem. This is difficult to detect, since it is often done on-site immediately after the rough is mined, and before the gemstone undergoes other treatment. This controversial treatment primarily involves corundums coming out of Asia.

So is this legal? Well, yes and no. The Gemological Institute of America (GIA) and stone suppliers have rejected the use of silica and aluminum, since these materials do not contribute in any other way to the enhancement of the gem. However, new policies have caused industry watchdogs to take a second look at borax, which scientists have shown improves the clarity of the stones. Today, dealers worldwide generally accept gems showing no residue of borax under a normal 10x jewelers loupe, which magnifies an object ten times its original size.

Heat

The most widely used method of gem treatment is the application of heat. An excellent example of this form of enhancement is the blue topaz, which almost always starts out white (or colorless). The longer it is heated, the darker the stone becomes. This is the reason blue topaz is seen in such a wide spectrum of colors, from the very pale sky blue to the almost sapphire-like London blue.

Other gems are heated to add needed clarity to a stone already colored by an invasion of impurities. These gems are heated in a rather primitive manner, to bring out their color and clarity. If the treatment is done correctly, no residue should be visible under a 10x jewelers loupe. The blue sapphire, a beautiful corundum, is an example of a gem commonly exposed to this process. But many other gems are enhanced in this manner. It may raise an eyebrow or two, but the fact is that nearly every aquamarine on the market today has been heated, in many instances to transform an unattractive stone in pale green or yellow into a beautiful, clear, icy blue color. The heat-treatment process is unquestionably the most widely used form of gem enhancement today.

Irradiation

Blue topaz and other gems are also color enhanced by irradiation—but do not panic: your finger, wrist, neck, or ears will not glow in the dark. The process uses only low levels of radioactivity, and the Nuclear Regulatory Commission (NRC) monitors the release of these gems. This is a practice that has been in place for many years, and I can assure you that not one case of radiation poisoning has ever been reported from wearing a blue topaz, or any other gem for that matter. In fact, microwave ovens often generate higher levels of radiation, and they are used by consumers every single day. Personally, however, I believe that someday the radiation levels for manufacturing processes like these will be so overregulated that the dark blue color we know today in the London blue topaz will be unaffordable or cease to exist entirely. (That's only my opinion. Feel free to reject it if you wish.) Other gems that are sometimes irradiated include the diamond, the aquamarine, and—perhaps surprisingly—the pearl.

Oiling

Treating gems with oils is also a popular method of color enhancement. Most emeralds seen in the world today are treated with an oil of one kind or another. In many instances grape-seed oil, which sometimes contains a green dye, is used. If the stone has white cracks common to the crystal, the dye is added to ensure a vivid consistency of color that will last for many years with proper care. Consumers should know that this oil does not last forever, but you can ensure the quality and longevity of your emeralds with a periodic light application of olive oil or another variety of cooking oil. Just get a clean, soft cloth and rub lightly; after waiting a few moments to let some of the oil be absorbed, wipe it off. If you do not want to be bothered doing it yourself, many jewelers will offer this service to their customers as a courtesy. This is an evolving area of gem enhancement, and new materials said to be more durable and long-lasting are in the development stages even now.

Consumer Resistance

If you are like most folks, your first reaction to all of this will be negative, causing you to reconsider and maybe take up butterfly collecting instead. If you're tempted at this point to say, "Well, I'm just not going to buy any gem that has been enhanced in any way," I can tell you firsthand that your resulting gem collection will be less significant than a blade of grass in a hurricane. This would be a pity, because in all honesty the beautiful effects that are so pleasing to the human eye would be impossible without these enhancements. As further evidence of the importance of gem treatments, do yourself a favor sometime and visit a rock and mineral show when one passes through your area. Take the time to look at the rough; inspect these "beauties" in their natural state. If you don't want to bother, I can save you the trip: you're going to find things that more closely resemble what is at the bottom of your gas grill. If it will make you feel any better, always remember that in many instances the color was always there, thanks to the miracles of Mother Nature; it just took a little help from people to bring it out.

It probably will not do you much good to ask the salesperson behind the counter how a particular gem is enhanced; by now, you probably know more about it than the clerk. If you are making a major purchase, determine the store's return policy, then take the item to an expert for evaluation. Remember, it's your money.

Disclosure

One of the most serious challenges facing the gem community these days is the issue of disclosure. Up to this point, little has been done to safeguard the industry, and ultimately the consumer. Until recently, manufacturers did not give this issue much thought, and often dealers just did not know enough about the enhancement of their gems. If the dealers didn't know, consumers had even less information to go by, and little chance to learn much about their purchases. Well, as Bob Dylan once said, "The times, they are a-changin'." Fueled by knowledge from nationwide television shopping networks, TV news magazines, and various publications, people now want to learn even more.

So why has it taken so long for this information to reach the consumer? There are several reasons. First, much of what passes for gem treatment is done in remote places of the world, far removed from the glitzy shopping malls you frequent near your home. Second, given the many steps of mining, excavating, treating, cutting, exporting, importing, and retailing, this information can get lost in the shuffle. Finally, fears of consumer rejection of anything that has been treated may make informed sources dispense this information only if a potential buyer requests it.

I may not be able to speak for everyone, but in this instance I can speak for myself. During a candid interview with David Pillow, director of quality assurance at TV's Home Shopping Network, I learned a lot about quality control. David searches for any trace enhancement, fillings, dyes, or irregular striations (bands) resulting from the treatment process. If any are detected, the product is rejected until he can confirm his findings or get proper disclosure from the vendor. Depending on the outcome of the investigation, the item is either placed in inventory or returned to the vendor. We take the utmost pride in our strict quality standards; still, we are aware, as you now already know, that nearly every gem found worldwide is treated or enhanced in one way or another. The important thing to remember is that these stones, when treated and cared for properly, will offer the consumer countless years of enjoyment.

So how can the consumer avoid the pitfalls of unethical gem treatments? For starters, deal with established retailers who have lived up to your expectations in the past. (First-time buyers should seek recommendations from trusted friends.) This alone is not a safeguard, because even the trusted retailer may unwittingly be duped by a dishonest vendor. Your best defense will be your own education. Since the government is not charged with the duty of educating the everyday gem consumer, continue this process yourself. Keep up with the latest industry advances through books, magazines, and newspapers. For your convenience, I have included a list of recommended reading materials near the end of this book. For consumers who are computer friendly, the Internet can also be an invaluable source of information.

Personally, I think the entire industry is overreacting to this situation. What it all boils down to is—as usual—sales. Exporters are afraid to tell importers, dealers are afraid to disclose information to the retailer, and the retailer passes the paranoia on to the consumer. Sorry, but I think the consumer is more intelligent than that. We now all know that gem rough is treated to enhance the beauty of the gem, and that this has been so forever. It should not deter you from making purchases in the same manner in which you have in the past. Gem treatment is simply a part of life, and if you deal with reputable sources, your problems will be minimal. So relax and enjoy your treasures.

The Government's View

After much debate, the Federal Trade Commission recently declared that any treatment requiring special care by the consumer must be disclosed. For example, if the gem under consideration has been dyed, and the dye will run if the consumer cleans it in a certain manner, the dealer must disclose this fact. This ruling was welcomed by consumer advocates and regarded as a victory for gem lovers everywhere.

However, all is not rosy. The industry also declared a victory when the FTC agreed that stone treatments generally regarded as permanent (such as heating, irradiating, and oiling) do not have to be disclosed, since consumer awareness of these practices has become increasingly apparent. Although no specific groups were given the credit for increasing consumer awareness, I can't resist the chance to give the home-shopping channels a gold star for the role they have played in educating the everyday gem consumer nationwide.

Chapter 5

Man-Made Gemstones

Synthetics and Simulants

Today's modern technology has enabled scientists to duplicate natural gems in a laboratory environment. The term *synthetic* should be confined to a gem that has been manufactured so that its composition is virtually the same as that of its natural counterpart. This is what separates the synthetic, or created stone, from the *simulant,* or imitation. (*Simulant* is a word in the jewelry trade, shorthand for "simulated gemstone.") For example, a natural ruby is a member of the gemstone family, or mineral group, known as corundum; thus, in order for a stone to carry the label of synthetic ruby, it too must have essentially the same makeup as natural corundum. This assures the consumer that the stone has virtually the same optical, physical, and chemical properties as the natural ruby. If the stone in question were red glass instead, it would have a totally different composition, with nothing in common with a real ruby other than its color, and would therefore fall into the category of a simulant.

Some simulants look more like the real thing than others, and they can offer attractive choices at a more affordable cost than most synthetics. An excellent example is the ever popular colorless cubic zirconia. Although its properties differ from those of a diamond, this "fabulous fake" is so durable that manufacturers cut and polish these stones in much the same manner as they would a diamond. Indeed, high-quality stones of believable size and set in fashionable mountings can look like a diamond even to an experienced jeweler. But when such a stone is put to the test, it is quite simple for experts to identify its composition and structure.

The case of colored cubic zirconia is a little different. Like the colorless variety, the colored kind does not duplicate the composition of any colored gem;

unlike it, the properties of colored cubic zirconia are usually so different from those of any natural colored gem that it seldom looks like the real thing. It is for this reason that you may have heard someone describe it by using an expression such as "too perfect" or "glasslike" in appearance. If you desire some of the effects of a high-quality colored gem, cubic zirconia offers an attractive and affordable alternative. But if you want something that really looks authentic, then a synthetic or lab-created stone will be a better choice, particularly in the case of colored gems.

Sometimes simulants are created to improve upon nature. For instance, consider again the natural kyanite. It possesses a beautiful and distinctive blue color; however, this stone is too soft to be a good choice for jewelry, because it would scratch and wear too easily. An imitation kyanite was created by making a synthetic quartz in a color similar to kyanite. This resulted in a very attractive gem for jewelry lovers—and much more wearable than the real stone.

The lab-created gemstone offers consumers many alternatives to the natural gem, and the choice is one of personal preference. Do you sacrifice some of the properties of clarity and color for the pleasure of owning a natural gem, or do you go with the "million-dollar look" that the man-made gem delivers? If you are like most consumers, you will probably end up with a little of each.

Doublets and Triplets

Other interesting and affordable alternatives to synthetics and simulants are doublets and triplets. (A term sometimes used interchangeably with the triplet is *soudé,* pronounced "sue-day," which is making some headway in the retail marketplace; some people think it sounds more romantic.) The technology to make doublets and triplets has existed for many years, and they often continue to be encountered at retail today. It may be helpful to understand the composition of each.

Although the development of the affordable synthetic gem has overtaken the doublet in popularity, the latter still has its place in today's market. There are two varieties of doublets encountered at retail: "false" doublets and "true" doublets. A false doublet is a creation that consists of two layers of a colorless material joined together with an appropriately colored adhesive. Although the stone most commonly imitated is the emerald, it is easy to see how a manufacturer can create a false doublet as a ruby, sapphire, or virtually any other colored gem merely by changing the color of the adhesive. False doublets are sometimes encountered today by consumers shopping for antique pieces of jewelry.

A true doublet is made by assembling two pieces of genuine gem material with an appropriately colored glue. For example, a manufacturer can create a true emerald doublet by fusing together two layers of pale green beryl with an adhesive of an emerald green color. Others are formed through color combinations of the same gem—for example, two different colors of spinel, layered to provide an interesting and sometimes convincing effect.

Another such "gem" found at retail is the black opal doublet, and more commonly today the black opal triplet. The black opal doublet usually consists of a slice of precious white opal and a second layer that may be either a lower-grade gem or another material entirely. The mysterious allure of the black opal is achieved by adding a black adhesive, which holds the layers in place. The black opal triplet is essentially the same product, with an additional layer that more often than not is of colorless quartz. The colorless quartz adds protection and durability to the creation, and may in fact intensify the play of colors beneath it.

Occasionally a product will come along that combines the technology of the doublet with that of the synthetic or created gem. One such variety gaining in popularity is called a soudé tanzanite, consisting of two layers of a glasslike material fused together by a blue adhesive. Although the colorless layers can be made of just about any material you can name, the pieces I have personally encountered were made of fused synthetic spinel.

There is absolutely nothing wrong with purchasing a doublet or triplet, for along with cost savings and practicality they offer the consumer the opportunity to own a beautiful piece of jewelry. Be aware, however, that a high-quality creation can easily fool the untrained eye, and these stones have been sold as genuine to the unsuspecting consumer. To avoid becoming a victim, keep in mind the selling price of the supposed gem under consideration; if it seems too good to be true, be careful! For example, if you are shown a rare and precious Australian black opal of several carats and its cost is anything less than that of a four-year degree at Harvard, there's a pretty good chance that it's a doublet or a triplet. Fortunately, in the case of the opal doublet or triplet, the layering of white and black is easily seen by examining the piece from the side. Be suspicious of settings that feature fully enclosed stones, for these are much more difficult to detect. Dealing with a reputable, established source is always your best choice, although unscrupulous suppliers have duped even the most honest retailer from time to time. If you suspect that a stone sold to you as genuine may instead be a doublet or triplet, a trained gemologist or jeweler should be able to detect an impostor from its genuine, costlier counterpart.

Chapter 6

Things You Should Know Before You Buy

In an effort to become the best jewelry shopper you can be, you first must master a few basic concepts. Understanding them all will prepare you for your shopping experience. By dropping a line like "Excuse me, but what is the specific gravity of that gemstone?" you will strike terror into the hearts of every retail gem clerk from Anchorage to Zimbabwe. Arm yourself with knowledge, for it's your best weapon against misinformation, whether accidental or deliberate.

Carat Weight and Specific Gravity

If you've ever watched one of my televised gemstone shows, I'm sure you've heard this before: "Always remember, the carat is the *weight*, not the *size*." So if you think I'm going to remind you again, you're absolutely correct. Only this time I'm going to explain it in a little greater detail. It may be boring, but it is nonetheless an essential concept in "Basic Jewelry 101."

When it comes to weight, a carat is an extremely tiny amount, just one-fifth of a gram—and it takes nearly thirty grams (thus, nearly 150 carats) just to equal one ounce. (That's good news for people who have to walk around with all those carats on their finger, wrist, or neck.) The weight of a particular stone is also expressed in specific gravity, which compares the stone's weight to the weight of the same volume of water. To keep it simple here, suffice it to say that a test is performed in which a stone is placed into a volume of water and then given a relative rating expressing its density. (In the gem trade, the word *density*—which used to refer only to a unit of measurement, such as four hundred milligrams per cubic centimeter or two ounces per cubic foot—now is used commonly to mean *specific gravity,* referring to a ratio of two weights expressed as a precise number, such as 3.20.) It is also important to remember at this point that a stone's density (or specific gravity) is not related to its hardness (see the next section); for now, let's just say that specific gravity is directly related to weight, while hardness refers to a gem's ability to resist scratching. Some gems, such as the relatively fragile malachite, have a hardness rating of only 3.50–4.00, which puts them quite low on that scale. However, this same brittle stone has a specific gravity (or density) of 3.80, which puts it above most other gemstones, including the diamond.

One of the most widely misunderstood concepts among consumers involves the relative size of the diamond and the cubic zirconia. Which is larger, you ask? The answer may surprise you. The cubic zirconia has a much higher specific gravity, a hefty rating of 5.80, compared with the diamond's density of 3.52. This means that the cubic zirconia weighs more than the diamond, and therefore one with the *same* weight in carats as the diamond will be *smaller* in size.

If this still sounds confusing, consider the following example. Let's say you had some work to do around the house, and you needed to buy fifty pounds of cement mix and fifty pounds of grass seed. Which bag will be smaller, the cement or the grass seed? If you said the bag of cement, you are 100 percent correct. Since cement weighs more than grass seed, it will require a smaller bag to contain fifty pounds of cement than to hold fifty pounds of grass seed.

OK, let's go back to jewelry for a quiz to see if you fully understand the concept. You have decided to purchase a ring for your mother, and you know she hinted at a solitaire of about two carats. You also know she loves the color blue. Your choices include the blue topaz, with a specific gravity of 3.54; the aquamarine, at 2.69; or the sapphire, with a specific gravity of 4.00. Assuming you go with the two-carat stone, which one will appear to be largest in size: *(a)* the blue topaz, *(b)* the aquamarine, or *(c)* the sapphire? If you selected the aquamarine, great going! Since the aquamarine is lighter in weight (density) than the blue topaz or the sapphire, to equal two carats it must be larger in size than the others. The blue topaz would obviously be the second largest, and the sapphire the smallest. For the specific gravity of other gems, see "Key Properties of Gems and Minerals" in chapter 14.

Before leaving this section entirely, I want to clarify another point. Some guy, probably someone with an ax to grind against jewelry consumers worldwide, thought it might be funny to throw another type of carat—this one spelled *karat*—into the mix. This was obviously a cruel, inhumane trick, devised by the sort of person whose idea of fun is developing a lengthy research project on the history of gum. Fortunately for the rest of us, this joker obviously flunked remedial spelling, which is one way to tell the two words apart. The other thing you need to remember is that one has absolutely nothing to do with the other: *karat* has to do with the quality of gold, not the weight of gemstones. Luckily, precious metal is sometimes identified in the jewelry world specifically as "karat gold," which helps. (Just to add to the confusion, in some European countries, the abbreviation *ct.* for gemstone weight is used interchangeably with *ct.* for pure gold content.)

For the most part, as long as you keep your gold lingo in a separate corner of your brain from your gemstone lingo, you'll be OK. For some reason, I have all the confidence in the world in the female section of the crowd, but I can't say the same for the men. Hey, guys! This is your chance to prove me wrong!

The Mohs' Hardness Scale

Hardness, as I previously mentioned, is loosely defined as the ability of a particular stone to resist scratching. For years the jewelry trade has relied on a unit of measure, called the Mohs' hardness scale, to define the fragile nature of one stone compared to that of another. This benchmark of the jewelry industry was developed by the German gem expert Friedrich Mohs way back in 1812, and experts still use it to this day. The scale is expressed in numerical values of one through ten, with one being the most easily scratched, and ten the most scratch-resistant. What Mohs did was to select ten common minerals and rate them according to this "sliding scale of scratchability": (1) talc, (2) gypsum, (3) calcite, (4) fluorite, (5) apatite, (6) orthoclase, (7) quartz, (8) topaz, (9) corundum, and (10) diamond.

Mohs found that each of these minerals can scratch those with a smaller number, and in turn each can be scratched by any of those with a larger number. All other gemstones can be rated in the same way, by comparing their hardness with those of the ten selected minerals. The hardness level of a particular gemstone is helpful in understanding which stones in general will lend themselves to normal daily wear, and which ones should be worn while at leisure. For example, if you are a particularly active person, a ruby, with a hardness of 9.00, would be a better choice for everyday wear than the more fragile opal, with a hardness rating of 6.00. "Key Properties of Gems and Minerals" in chapter 14 lists the hardness of other gems.

Toughness

Knowing the hardness of a particular gemstone can lead you to many helpful hints on appropriate care and cleaning methods; likewise, it can assist you in selecting that special gem for that special purpose. Hardness is certainly an important factor when making a decision, but there is another, similar consideration that most casual gem buffs often overlook: toughness.

Unlike hardness, which indicates a particular stone's ability to resist scratching, toughness defines its resistance to breaking. This is a very important factor in considering durability, and a few surprises are in store for the everyday gem lover. For instance, the diamond is the hardest gem known, but its toughness rating is only "good." Jade, on the other hand, is easily scratched yet prized as a beautiful carving stone, categorized as being "extreme" in toughness. In fact, it has been said that the structure of jade makes it "stronger than steel." Before purchasing a stone, you should consider its durability in relation to its intended use; while this should not be the sole factor, it is important enough to bear at least some weight in the decision-making process. This consideration is most important when selecting a bracelet, and least important when considering a purchase of earrings. Keep in mind that any gem, if cared for and handled properly, will last the wearer for many, many years.

Understanding the Refractive Index

In simplified terms, the refractive index (RI in the gem trade) measures the speed and angle of light as it passes through a gemstone (light moves more slowly through stones than through the air, and its exact speed depends on the type of stone). These readings are measured by a trained jeweler or gemologist using a piece of equipment called a refractometer. Certain gems will also split the light ray in two; these gems, which include beryl, corundum, peridot, topaz, and tourmaline, are said to be doubly refractive. It may surprise you that the diamond, with all its sparkle, is merely singly refractive. Because many gems look alike, a thorough understanding of the refractive index is a most useful tool in the identification of gems.

I guess all of this stuff is necessary for technical people with dour dispositions and white lab coats, but personally I'd be willing to bet it's not high on the average consumer's list. As a matter of fact, I did some research on the matter myself, and I found that although the refractive-index table (see "Key Properties of Gems and Minerals" in chapter 14) was developed hundreds of years ago, since then only sixteen people have ever been able to completely understand what it means, and all of them either are dead or live in cold-climate countries with names I can't pronounce. Still, if you get nothing else out of this discussion, now you can walk into the jewelry department of some overpriced department store and ask the clerk to quote you the refractive index of the blue topaz that's on display. The answer may be a source of conversation for some time.

This concludes the high-tech discussion on refractive index. If you have any further questions on this subject, don't ask me, because I've already admitted I'm clueless; in addition, I don't have a dour disposition and I don't even own a lab coat. You could try contacting Willebrord Snell, the Dutch mathematician who pioneered this formula. However, since the concept was developed in 1621, it's not likely you'll be hearing from him anytime soon.

How Color Happens

The eye sees color through a unique interaction with the brain. In most cases, a gem's colors are produced primarily through the absorption of light hitting it.

Light is made up of different colors (each of which has a different wavelength), and the colors that are not absorbed but instead pass through the stone are the ones you see. For example, a ruby absorbs all colors other than red, so when you look at it you see a red ruby. Gems get their color, or colors, mainly from so-called impurities that invade the rough. Among these inclusions are iron, manganese, chromium, vanadium, cesium, and lithium. Most gems then need a helping hand to bring any colors to the surface. It seems ironic to me that science calls these coloring agents "impurities," a word that has a negative connotation; after all, these same so-called impurities are responsible for the explosion of colors that fill the rainbow of the gem world and provide the consumer with so many choices.

The Importance of Color in the Evaluation of Gems

Color plays a significant role when someone is establishing the value of a gem. Two gems from the same mineral group might not both be thought of as gem grade or gem quality should their color range be far apart. For example, one emerald could be a rich green beryl, while another could be almost totally devoid of color. It's obvious that the first one would be more than welcome in the gem world, while the second may not make the arbitrary grade thanks to its lack of color.

When evaluating a jewelry item for purchase, remember that consistency of color within a single stone is equally important. Modern methods of treatment can cure many ills, but a lack of color consistency can still be obvious. Check to see that the stone has no so-called dead spots; the color should be true when you look at the piece from any angle. If the piece of jewelry being considered consists of more than one stone, as in the case of a bracelet or a cluster ring, compare the stones to make certain they are a perfect or near-perfect match. Consistency of color is not only important in the selection of colored gems, but vital in evaluating pearls as well. Of course, color is a matter of choice. Whatever color you select, however, one fact can't be disputed: its consistency is crucial in determining its value.

Some of the Many Varieties of Shapes

Oval *Heart* *Trilliant*

Pear *Marquise* *Baguette*

Two terms that consumers often misunderstand are *cut* and *shape*. Cut is used to describe the way a stone is faceted—which refers to the many small flat surfaces, angled to one another, that an expert produces on a stone to increase its beauty—while shape has to do with the actual form of the stone. Some of the shapes most commonly encountered on a daily basis include round, square, oval, pear, heart, marquise (elliptical with pointed ends), trilliant (triangular with additional facets on its underside), and baguette (a long, narrow rectangle). All of these, other than round, are generally classified as fancy shapes. However, advances in modern technology have made the world of gemstone shapes nearly limitless today. Although your choice will probably be guided by your passion, other considerations should come into play as well. Consider the shape of your finger, for example, since certain shapes will look more flattering than others on your hand. Remember, too, that if you plan to wear the ring in conjunction with another, the shape you choose may prove to be impractical.

How to Use a Jewelers Loupe

A jewelers loupe (pronounced "loop") is a specialized magnifying glass with a power of 10x (ten times). This means that an object you are examining when you look through the glass appears ten times closer—and larger—than through the naked eye. These loupes can be found in optical shops and some jewelry and department stores. Although there may be more than one to choose from, your loupe should cost no more than $25 to $50.

There is a reason that the loupe sees an object with a magnification of ten times, rather than twenty or thirty times. It is an industry standard in the gem world that gems be graded under this power when a person is looking for flaws; if a flaw is not apparent at this magnification, it is considered nonexistent. A jewelers loupe can be a friendly asset when checking for flaws that would not be apparent when viewed with the naked eye. The case that holds the loupe should be black, so as not to interfere with the color of the stone when you are looking at it.

When examining jewelry through a loupe, hold the lens as close to your eye as is practical for you, and slowly bring the gem up to the loupe. With some practice, you will become comfortable with your loupe in no time. What you must then keep in mind is that few, if any, stones you will ever own will appear to be completely flawless through a loupe. My favorite motto is "You get what you pay for." This is to say that if you are looking at a $200 ruby, the chances of its looking under the loupe like one that is worth thousands of dollars are remote at best. The trick is to remember that if you got a good value and the ring does not show any defects to the naked eye, you should be happy.

How to Detect a Properly Cut Gemstone

Checking to see if a gemstone is cut in proper proportion is extremely important, because two gems of otherwise equal quality can differ greatly in value if one is improperly cut. Indeed, in the case of the diamond, a properly cut stone can be worth twice as much as a gem of similar quality that is improperly cut. In addition, poor faceting can actually weaken the stone, making it more likely to chip or crack from even the most insignificant amount of contact with another hard surface.

There are certain things even the average jewelry consumer can do in order to distinguish a nicely proportioned stone from a poorly cut one. Begin with the table, or flat facet at the top, of the stone: check the faceting for proper proportions all around (a sense that the stone seems to be in balance). Look for depth of color, if applicable, and the presence of dead spots (inconsistent color) that may be clearly visible to the naked eye when the stone is viewed from various angles. Check, too, the depth of the cut (how sharp an angle there is in the facets from the top of the gem to the nearest sides), by examining the stone from all sides and the bottom. If the cut of the gem in question appears to be too shallow, you may be looking at a stone that has been improperly cut. One that has been cut across its table with little depth can result in a windfall of profit for a gem dealer, as untold carats are saved by not cutting the gem to the proper depth. Although that stone may seem larger to the inexperienced consumer, in actuality the value of the piece may be far lower than expected. This practice is sometimes known as shaving.

A poorly cut stone can reveal other detectable imperfections. In stones with too shallow a cut, you can often see a reflection of the girdle, or widest part of the stone, through the stone's table. In diamonds and occasionally in colored stones, this is referred to as a fisheye. In colored stones, shallow cuts also can create spots where you can actually look into the interior of the stone. These spots are sometimes known as windows. Conversely, if you have a stone that looks dead, or too dark, it is more than likely that the stone is cut too deeply, and it is actually leaking light. This effect is known as a nailhead.

Evaluating Settings from Every Angle

Choosing a setting that is appealing to you is a personal matter, and like choice of color it should be left up to the individual consumer. However, there are important points to consider before spending your hard-earned dollars.

Prong

There are a number of different ways to set a gemstone into a piece of jewelry, the prong setting (with tiny projections to hold the gem in place) being the one most frequently encountered. I have heard it said that six-prong settings are better than four-prong settings, to which I say it all depends on what you mean by "better." Certainly, a six-prong mounting will offer additional stability and reduce the likelihood that a stone will be lost from the setting, so I guess in that case it is indeed better. But some consumers feel that the extra two gold prongs can detract from the overall appearance of the piece. Personally, I think each ring, each stone, and each setting has to be evaluated, with personal priorities placed on looks versus security. With a large colored stone, for example, the addition of two more prongs will most likely have little, if any, effect on the aesthetics of the ring. The same may not be true of a smaller colorless stone like a diamond solitaire, or a cluster ring consisting of many small stones. No matter which you choose, when worn under normal conditions both are regarded as quite durable and should last a lifetime with proper care.

Another point worth some consideration is the color of the prongs. Many manufacturers will use a rhodium tip when setting a colorless stone such as a diamond in a prong setting. The rhodium tips will make the prongs less visible against a clear stone; further, rhodium is a brilliant white metal that will increase the brightness of the setting. If brilliance is of paramount importance to you, this type of prong setting is most likely going to be your best choice.

It should be noted that the prong setting also has certain negative aspects. If you wear silk blouses or certain other varieties of long-sleeved clothing, such as sweaters, be aware that a prong setting does tend to snag on occasion, which could be important—particularly in the case of a bracelet. In addition, if you perform work that requires a lot of movement of hands, wrists, or fingers, there is an increased threat of losing a stone out of the setting if you carelessly strike the piece against a hard surface.

Channel

On the other hand, channel settings (in which the stone is set into the grooves of two walls) are extremely secure and offer a wonderful front-face or face-up view of the gems. Channel-set gems are very popular and should be considered when selecting wedding bands; in fact, many retailers today offer wedding sets consisting of a prong-set diamond solitaire and an interlocking wedding ring with channel-set gems to complete the look. The only real drawback to the channel setting is a loss of sparkle. Since the stones are recessed into the ring rather than raised in a high mounting, the channel setting does not offer the same sparkle as the higher prong setting does.

Bezel

For security and style, most men prefer their jewelry in either a channel or a bezel setting. Bezel settings feature a stone or series of stones surrounded by a continuous framework of precious metal (such as gold, silver, or platinum), rather than a series of prongs. Bezel settings offer a level of security and protection unmatched by any other. This is quite important when incorporating more fragile gems, such as pearls and opals. Bezels also make a gem seem larger than it actually is, and they also may hide certain chips, cracks, or other imperfections that would be clearly visible in a different type of setting. Colored stones tend to darken somewhat in bezels (because the stone is set deeply and less light may reach it through the setting), and this can enhance—or sometimes detract from—the look of the piece.

Bezel settings were used almost exclusively in ancient times, not only in articles of personal adornment but also as a means of placing gems into swords, shields, and amulets. Depending on the designer of the piece, a bezel setting can take on an old-world, romantic appearance (for example, gems set with marcasite) or an ultramodern sleek look, usually incorporating unusual gems cut *en cabochon*.

Unfortunately, because it sometimes encompasses more of the stone, a bezel setting can result in even greater loss of sparkle than a channel setting. Bezel settings nonetheless have their place in the world, particularly when the stone is fragile; in addition, they best accommodate any semiopaque gem, such as jade, onyx, or agate. As a rule, bezels are also the most popular choice for gems showing a special property, such as a star or a cat's-eye effect. Keep in mind that bezel settings can command a higher dollar at retail, as a result of the extra amount of precious metal used in this type of setting. Unquestionably, though, the bezel setting does offer a look all its own. Before making your purchase, consider this greater cost together with the quality of the gem involved, to be certain the potential purchase is still a good value.

Less common, but still somewhat prevalent on the market today, is the pavé (pronounced "pah-vay") setting. This type of setting gives the illusion of a "pavement" of gems placed very close together, which no doubt accounts for its originally French name. In order for a setting to be called pavé, one prong must touch three or more stones in the setting. If not, the gems are considered set too widely for a pavé setting.

Pavé

In order to make gems affordable to as many consumers as possible, manufacturers will sometimes introduce additional amounts of rhodium or other white metal into settings. Sometimes known as an illusion setting, this may incorporate a few smaller, less-costly diamonds, or in certain cases no diamonds at all. The white metal may be diamond-cut to give the appearance of diamonds, or to further enhance the smaller gems scattered throughout the setting. More often than not, this effect is incorporated with colored-gem solitaires, as an accent to the main focal point of the setting. Because it makes jewelry more affordable for a wider group of people, this effect should not be thought of in a negative light. However, the salesperson should clearly communicate to the customer when this is used. Life being what it is these days, this is unfortunately not always the case.

The consumer may encounter a number of other settings in today's anything-goes marketplace. Two of the most common of these are the bar setting, which is sometimes confused with the channel setting, and the bead setting, which is sometimes confused with the prong setting. By no means does it end here. Designers may incorporate two or more different settings (such as channels and prongs, for example), or introduce settings all their own. Whatever setting is used, in the end the choice is yours. In this case, beauty truly is in the eye of the beholder.

Finally, a word of advice. Certain settings will occasionally incorporate an adhesive into the mounting; this is sometimes seen when evaluating marcasite. About the only advantage of glue that comes immediately to mind is the elimination of prongs, which may be of top priority to you. Personally, I detest the concept of holding a gemstone—or even a man-made simulant—in place with glue.

After all, would you go around bragging to your relatives and friends that your stones are set with the highest-quality glue available? Somehow, I don't think so.

The Importance of the Clasp

Another consideration when shopping for jewelry should be the clasp, because a poor-quality one can pose problems later on, resulting in the loss of a valuable possession. There are many different varieties on the market, and no one clasp works best for every piece.

Rope chain, for example, is usually found with a barrel clasp, as the round shape of the clasp seems to complement the round shape of the rope. To make it work, insert the elongated tongue into the barrel section of the clasp and then fold the extra security hook over the top. To remove, merely reverse the process, first unfolding the extra security hook and then depressing the safety-release lever. Be careful when opening this clasp; pull gently and depress the tongue with just enough force to get it open. Too much pressure may cause a loss of tension.

Barrel clasp

Although in general it may be the least expensive clasp, the spring-ring clasp is still the most common type and performs its duty quite well under normal circumstances. At one time many people shunned it as being too weak, but lately large spring-ring clasps known as toggle clasps have brought this kind back into prominence. This works particularly well in the case of smaller pieces. To operate the spring ring, merely slide the lever on the larger side open and then insert the smaller side into the opening. To remove, simply reverse the process.

Spring-ring clasp

The lobster-claw (or fishhook) clasp is often seen in the herringbone chain and most varieties of link chain. Pulling down on a lever opens the clasp enough so you can insert the other end and fasten it securely in place. To remove, simply reverse the process. Today, many consumers spend extra dollars having spring-ring clasps retrofitted as the lobster-claw kind.

Lobster-claw clasp

If cost is no object, the inserting-box or tongue-in-groove clasp is generally accepted as the easiest to use and most secure clasp of all. This clasp opens when you push down on a plunger or safety-release lever while gently pulling one side away from the other. To close, simply reverse the process. When using this type of clasp, you should hear the tongue actually click into place to indicate that it is closed correctly. If no click is heard or felt, it is probably because the tongue has been pushed too far down and will not react when set into the opening. To remedy this situation, all you have to do is use a thin object (such as a tiny screwdriver) and carefully spread the tongue apart.

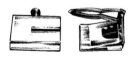

Inserting-box clasp

Besides these main types of clasps, there are myriad other varieties one will encounter, from delicate yet ornate filigree pieces on strands of cultured pearls to large customized creations designed to be shown rather than hidden. No matter which clasp is involved, remember that many have delicate springs, levers, and tongues that someday may need to be replaced. After making a purchase, check the clasp often to confirm that it still provides the security it did originally. Don't wait until you find out the hard way that the clasp needed to be replaced.

In summary, the choice of a clasp is usually pretty much a matter of personal preference, although there may be certain practical factors as well, and not just cost. Among other things, a person with arthritic hands, or a lady with long fingernails, may not be able to operate certain varieties. To these consumers, the decision has less to do with looks than with necessity and practicality. Therefore, select the clasp that most suits your needs as well as the design of the piece, and always check it to be certain it is working properly. Personally, if a clasp seems cumbersome on the first try, I would either make another selection or plan to have it replaced at the first opportunity. Don't be afraid to voice any concerns to your retailer; if the piece in question is brand-new, ask the retailer to replace the existing clasp with one that is more suitable for you. Whatever the circumstances, never underestimate the importance of the clasp.

The Value of an Appraisal

Some consumers believe there are two kinds of appraisals: oral and written. In fact, an oral appraisal is nothing more than an educated guess as to the value of the item. Although it may be of some comfort and give you an idea if your purchase was a good value, insurance companies do not accept oral appraisals as evidence of worth.

If the piece of jewelry you have purchased is of significant value, it would be well worth investing in the nominal costs of having the article appraised by a gemologist certified by the Gemological Institute of America. In many instances you will be asked to leave the article with the dealer, as some appraisals can be quite involved and take hours to develop. If this is the case, always get a receipt from the dealer before leaving the store. Determine who is going to do the appraisal, and where it will be done. Personally, I would deal only with someone who is going to examine the piece on-site, and not with anyone I am unfamiliar with. (First-time buyers should ask friends for a recommendation.) I don't mean to sound so skeptical, but unfortunately that's the way of the world today.

Always remember, however, that appraisals are by their very nature subjective, and they can vary greatly not only from state to state and from city to city but even from dealer to dealer. If you are at all unhappy with the first analysis, it might be a good idea to get a second opinion, in the form of an appraisal from another source, before accepting that you made a disappointing buy or even got stuck with a poor-quality gem. Remember, too, that a trained professional can also appraise the value of jewelry items without gemstones (gold chains, for example).

Upon completion of the formal, written appraisal, check with your insurance provider to see just how much of your beautiful jewelry is actually covered under your policy; more often than not, the coverage is in no way sufficient to cover a loss entirely. If you find yourself in this situation, you may want to consider adding a rider to your present homeowners' or renters' policy, for additional coverage. Inquire as to the costs, deductibles, and other important factors before deciding to increase your coverage. Insurance companies usually require proof of ownership, so make sure a close-up color photograph is included in the final appraisal. Keep in mind that jewelry may appreciate in value over time, so it would be a good practice to have your high-ticket items reappraised at certain intervals. It is also a good idea to photograph or videotape all of your treasures, and keep the photos or videos, along with the items' appraisals and sales receipts, in a secure location, preferably in a fireproof container. It never hurts to be protected.

Inspection, Wear and Tear, Repair

I advise you to check your jewelry purchases from time to time for telltale signs of wear and tear. Prongs sometimes have ways of loosening and clasps sometimes give way. Rings made of karat gold can split when subjected to extensive use. If you suspect that a piece of jewelry you own may be in need of repair, find a reputable retailer that offers this service. When dropping off a piece of jewelry to be repaired, be certain to get a receipt, with a clear description of the object. As in the case of an appraisal, ask the dealer if the repairs will be done on-site or if the piece will be sent out instead. If it is to be repaired at another location, find out exactly where. Inquire, too, about how it will be shipped, and whether your piece is insured during its time away from you. In addition, determine the anticipated method of repair, and get an estimate in writing.

Inspection and protection of your gems are absolutely vital. Those of us who have ever lost a stone or a piece of jewelry know that it is not a pleasant thing. My wife, Linda, lost the first Valentine's Day gift I ever gave her: a heart-shaped filigree bracelet in 14-karat gold. Although I will never know for sure, I suspect that the delicate wire gave way at the clasp. It wasn't so much the cost, which was less than $200, but as you can probably imagine, it was irreplaceable because of its sentimental value.

Chapter 7

$\mathcal{S}$electing a Retailer

$\mathcal{W}$hom Can You Trust?

The proliferation of man-made gems has made shopping for gemstones more confusing than ever for the unwary consumer. Advances in technology are making it increasingly difficult to tell whether a stone is man-made or natural. Reputable dealers will identify the lab-grown gemstone as such, and today's sophisticated methods of evaluation allow the experienced gemologist to detect even the subtlest of differences. My advice is to deal only with a well-established retailer who has a reputation for honesty. Especially shy away from sales offered by private individuals, unless you are personally acquainted with the seller.

Let common sense, not emotion, be your guide. Certainly cost is an important factor, but it should not be the only one when deciding what to buy and from whom. When evaluating costs, try to compare the asking price with the market demand for the gem and its availability. This will give you some indication of whether the cost is in line with what it should be or if you are being taken advantage of because of the difficulty of locating the gem. If you are buying an item in either genuine karat gold or sterling silver, be certain the precious-metal content is indicated somewhere on the piece you are considering.

If there is a guarantee with your purchase, make sure to get it in writing, and hold on to your sales receipt, just in case. Be certain the return policy of the retailer is clearly stated, and get that in writing too whenever possible. See, too, what special services the retailer offers to the consumer. This may include such things as free examinations, tightening of settings and clasps, and cleanings. Find out if the retailer offers other services, such as appraisals and ring-sizing options. Have the costs (or at least the estimated costs) of such services in writing before taking advantage of them.

The overwhelming majority of dealers you will encounter work for honest, reputable firms that place customer service and satisfaction at the center of their focus. Unfortunately, there is a small chance you may encounter a "consumer predator," well schooled in the art of deception. As in the case of most other things in life, this will be the exception rather than the rule. If you purchase your little treasures from a reputable person, a nationwide firm, or an establishment that you have been dealing with for years, you should have every reason to be 100 percent confident when making your choices. If anything unsatisfactory does occur, most likely it will be attributable to circumstance or human error. In many instances, a retailer will go above and beyond the stated length of warranty (within reason, of course) just to keep a loyal customer. Finally, a word of general caution: more often than not, if a deal seems just too good to be true, it usually is. Unless you have done your research and feel fully confident that the offer is legitimate, in the long run you will probably be better off to let it pass you by.

In chapter 14, "The Easy Reference Lists," you will find pointers to follow before taking the plunge, including questions you should ask the retailer—and yourself. Sometimes I find the retail marketplace to be as difficult to navigate as a room full of mirrors. I think you will find these lists to be an invaluable tool as you travel through the perplexing world of retail gems and jewelry. Do your homework, and do not hesitate to ask any questions that may arise. If the person you are dealing with seems unable to answer your questions, it is probably because he or she lacks the background training necessary to field them. No matter what, be sure you are well informed before making your decision; it may be the difference between a practical, smart purchase and a costly mistake.

$\mathcal{F}$rom Rough to Retail

Lovely finished jewelry starts out as crude, uncut stone called rough, which often has to be chiseled away from the mother rock in mines that are far underground—a risky business for the miners.

Whether the rough consists of large stones from a mine or tiny pebbles from alluvial deposits found in streams or other bodies of water, all of it has to be sorted according to variety, size, shape, and color, by experienced people who are often trained starting at a very young age.

Depending on size, shape, composition, and the direction of interior crystal formations, gemstones are then cut into workable material by everything from big circular saws to hammers and chisels to precise cutting tools.

Once cut down to the desired size and shape, each gem is sent to an expert for shaping, faceting, and other treatment to bring out its maximum appeal. After placement in a setting, the gem is ready to be moved to a retailer's display case, and ultimately into the possession of a consumer.

Chapter 8

Diamonds

As we have already seen, the line of demarcation between precious and semi-precious gems is pretty much a thing of the past. Still, I cannot just dump the noble and prestigious diamond into a general alphabetical pool of gemstones; it is worthy of its own separate category. Although the popularity of colored gems has narrowed the gap considerably, the diamond—which is the hardest of all minerals—remains the single most popular gem in the world. In this chapter, you will find out why.

How Diamonds Are Formed

Diamonds were created billions of years ago out of carbon deposits subjected to intense heat and pressure deep within the cavities of the Earth. It always struck me as ironic that the noble diamond is actually the simplest structure of all: crystallized carbon. Thanks to the volcanic activity that brings them to the surface, diamonds may be found in rock structures sometimes known as kimberlites, within vertical geological formations called pipes. The presence of certain other gemstones (peridot, for instance) sometimes announces the existence of these diamond deposits. Quite often, after years of weathering, diamonds removed from the mother rock have also ended up in the water of alluvial deposits. Naturally colored diamonds, which are rare, occur when certain impurities get into the carbon crystal.

Significant Sources of Supply

The hierarchy of diamond sources has shifted with the passing of time. Until the eighteenth century, the leading source of diamond production was India. Today the most important source of the diamond is the continent of Australia, home of the world's largest diamond mine, the Argyle mine. This one site accounts for approximately forty million carats of diamonds per year! The Argyle mine produces nearly the entire world's supply of pink diamonds, as well as a variety of other colored specimens, some commanding prices in excess of $1 million. Since about 80 percent of the diamonds consumed worldwide are earmarked for industrial use, it is worth noting here as well that Australia is the largest supplier of the industrial-grade diamond.

The first diamond ever discovered in Africa was found by a young child walking on the beach. Although he did not know it then, this discovery in the 1860s would fuel the beginning of South Africa's diamond era. South Africa became the world's premier source of diamonds, a position the country enjoyed for about a century. By now, South Africa has dropped to fifth in total diamond production. Other significant sources include Congo (formerly Zaire), Botswana, and parts of the former Soviet Union. Because of the availability of diamonds throughout Africa, that continent is known even today as the Diamond Continent, and Namibia (in southwestern Africa) produces the largest percentage of gem-quality diamonds in the world.

Some Important Diamonds Throughout the Ages

Personally, the subject I detested more than any other in my schoolboy days (or should that be "daze"?) was history. Every first period Mr. Stevens extolled the virtues of some event that happened before the discovery of rocks, while I struggled to keep even one eye open. He may have spoken in a monotone and looked like the guy who just missed the 5:30 bus to Chicago Heights, yet now I have to admit he was right: history can be important and interesting after all. What follows here is just a sample of the kinds of things I might have learned from Mr. Stevens—if only I could have stayed awake long enough to take a few notes. (I hope he's out there right now, reading these very words with satisfaction.) So please read on—and try not to fall asleep!

The world's largest cut diamond, weighing an enormous 545-plus carats after it was finished, was found in South Africa as recently as 1986. The large stone it came out of, which weighed 755 carats, was found to possess remarkable color characteristics, showing a beautiful gold with undertones of a fiery red, crystal-clear "soul." A few years and many tens of thousands of dollars later, this discovery produced the world's largest finished diamond; the spectacular gem was cut, appropriately enough, in an underground cavern near Johannesburg, South Africa. This site was selected by De Beers (the vast diamond empire) in order to guarantee the cutter an environment totally devoid of distractions, vibrations, and sound. In fact, they even painted the walls a soothing light shade of green, a color that is believed to relax the eyes and the body. (Readers with young children should take note of this fact!) The finished gem was first displayed for royalty in Thailand in 1995. A group of wealthy Thai investors and businessmen purchased the stone and presented it to their monarch in honor of the fiftieth anniversary of his reign. Since this is known around the world as the golden anniversary, the spectacular find became the Golden Jubilee diamond. Following a worldwide tour, the gem is going to come to rest in a scepter and be presented to Thailand's ruler.

Prior to the Golden Jubilee, the Star of Africa, a 530-carat wonder cut from the Cullinan stone (which weighed over 3,100 carats uncut—that's more than 620 grams, or well over a pound), held the title of the world's largest diamond. Today the Star of Africa is found in the Royal Scepter of Great Britain. Another beautiful gem is the 128.5-carat, canary yellow diamond known worldwide as the Tiffany, which was selected from the Kimberley mining area of South Africa. The legendary deep blue Hope Diamond, the single most popular object in the Smithsonian Institution, originated in India in the seventeenth century as an uncut rough weighing in excess of 112 carats (in modern carats, a very slightly altered standard that was adopted all over by 1913). It was reduced to a still-impressive 69 carats and sold to the king of France, then stolen more than a century later during the French Revolution, and after many years found recut and reduced to approximately 45 carats—still beautiful, but now surrounded by an enduring sense of mystery. Yet another important diamond is the 407-carat champagne gem known as the Golden Giant, which came out of the previously mentioned Argyle mine of Australia.

The diamond is the traditional birthstone for the month of April, and a nontraditional birthstone for February, July, and October. People who were born under the sign of either Cancer or Libra sometimes claim the diamond as their astral stone, and astrologers have linked it to the sun and Venus. The diamond is your stone of celebration for the tenth and sixtieth anniversaries.

Determining the Quality of a Diamond

To understand how wide the range of diamond quality is, all you have to do is pick up your local Sunday newspaper and scan the sales fliers inside. Prices will be literally all over the place, some of them even seeming to contradict others right within the same advertisement. For example, you may find a one-carat diamond tennis bracelet advertised at $199, another at $599, and still another at $1,299—all in the same ad, from the same store! If they're all one carat, then why is there such a disparity? The answer, of course, can be summed up in just one word: *quality*. As I've often said before, always remember: "You get what you pay for."

You may have already heard of the "four Cs of diamond grading": color, clarity, cut, and carat weight. These four factors combine to give a diamond its value. Although color is generally regarded as the most significant one, none of them should be overlooked when shopping for diamonds. Let's take a closer look at each, in order of importance.

Color

Of the four Cs of diamond grading, color is at the top of the list. The most sought-after diamonds in the world are those totally lacking any color. These diamonds can command tens of thousands of dollars or more per carat. A gemologist or jeweler can test a diamond for color by placing it on a black background while exposing it to ultraviolet light. This test yields the quality known as the fluorescence of the stone. Most diamonds fluoresce blue, but some show traces of yellow or brown. Unless it greatly detracts from the appearance of the stone in natural light, this underlying color has little effect on the value of a stone. If, however, this color is found to dominate, it can result in a devaluation during appraisal. Interestingly, the surface of highly fluorescent diamonds usually appears to be oily to the naked eye.

In the beginning, the letter *A* was so overused that the consumer was virtually at the mercy of the dealer: it seemed as if when one person described a stone as being of A quality, the next would say his was of AA quality, and so on. The Gemological Institute of America wanted to distance itself from this mess while helping consumers better understand what they were buying. The GIA concluded that by starting with the letter *D* (referring to diamonds closest to colorless) and ending with *Z,* it could classify a complete range of diamond colors. This system was ultimately accepted throughout the diamond industry and is still in use today.

Clarity

Because the majority of diamonds used in the manufacture of jewelry are quite clear, little can be observed by the naked eye. Diamonds are rated for clarity based on their appearance under 10x magnification through a jewelers loupe. (This tool can be an invaluable friend to the consumer evaluating a diamond.) This raises an interesting question: if you can't see it in the first place, then why worry about it? After all, when was the last time you showed someone your beautiful diamond, and he or she pulled out a jewelers loupe to look at it?

Nevertheless, grading for clarity is worth your attention—for a moment, anyway. Although many different grading systems indicate the clarity of diamonds, the one most widely accepted in the United States today is the system that the GIA developed. The scale begins with a grade of FL, followed by IF. These two grades are reserved for stones of flawless or nearly flawless nature. Unless you are a professional athlete or a rock star, you will probably not have to worry much about your diamond falling into one of these categories. Suffice it to say that most diamonds used in volume production today fall into the range from SI_2 (slightly included) to I_2 (included).

Cut

Consumers often confuse the terms *cut* and *shape*. Keep in mind that *cut* refers to the process of transforming gem rough into beautifully finished works of art, while *shape* refers to the actual form—round, rectangular, and so on—of the finished product.

Since the diamond is the hardest mineral known, it presents certain formidable challenges to the gem cutter. Basically, the diamond is cut by incorporating a diamond-tipped tool together with diamond dust. Because of the way the crystal forms, a diamond can fracture during this process when cut in certain directions, and it takes master cutters years to develop their craft. The quality and symmetry of the cut are significant factors in determining the value of any stone, and poorly cut diamonds can lose as much as 50 percent of their value.

The most popular cut of all is called the brilliant cut, which consists of a minimum of fifty-eight facets. Many shapes lend themselves to the brilliant cut, but by far the most popular is the round. The brilliant cut actually can mask some flaws and imperfections otherwise visible to the naked eye, such as carbon spots, bubbles, and inclusions. Some flaws can even be cut right out of the stone during the cutting process, thus enabling the consumer to have a diamond with fewer flaws and a better appearance. Although it will usually be apparent to the naked eye, if you are uncertain of the cut and the number of facets in the diamond, be certain to inquire. This is more difficult to ascertain when buying a diamond cluster or band ring, for these stones are generally quite small. In the case of a ring consisting of smaller stones, it is preferable that the diamonds be classified as *full cuts*. This term simply refers to a diamond that is a round, brilliant-cut gem.

The brilliant is arguably the most popular cut, but the consumer is likely to encounter many others on the market. One variety that has always been a classic in the gem world is the step cut, which features faceting like steps on the perimeter of the stone. A rectangular-shaped step-cut gem is often called an emerald cut. The step cut is frequently seen in combination with the brilliant cut. When faceted in this way, the rectangle is known as a radiant cut; when this same combination is found in a square-shaped gem, it is known as a princess cut. Another variety of radiant cut, less commonly seen but available, is the starburst, which provides the unusual effect that its name implies. Unlike the radiant and princess cuts, the starburst is often seen in round, oval, and occasionally marquise shapes, in addition to the rectangle and the square. Other cuts include the Swiss cut, which shows thirty-three facets, and the simplified single cut, with just seventeen facets or less. These two varieties are primarily seen in the manufacture of bands and clusters, sometimes in combination with lower-grade gems. Assuming all things are equal, Swiss cuts and single cuts should always cost less than full cuts or brilliants.

Brilliant

Emerald

Princess

Advances in technology have opened new windows of creative opportunity for the modern-day cutter, and the results can sometimes be staggering. Once limited to 57 or 58 facets, stones cut today with as many as 144 facets are beginning to take hold in the marketplace. Diamonds cut in this manner seem to almost explode with the eye-catching effect that we call sparkle. Although the temptation to purchase one of these wonders will be great, be aware that they do carry a hefty price tag, no doubt a result of their labor-intensive process. Truly, as this technology continues to advance, the diamond craftsperson will take full advantage of the creative opportunities that lie ahead, resulting in looks that would have been impossible just a few years ago.

When selecting a diamond, be certain that the cut of the stone is in direct proportion to the shape. Look into the stone faceup and check for dead spots visible to the naked eye. Make sure there is adequate depth by examining the stone from the side as well as the bottom, and above all don't be embarrassed to borrow a jewelers loupe for your inspection, if you don't already have one. Any reputable dealer will be happy to oblige.

Carat Weight

I have always found it curiously interesting to hear someone say, "Oh, you should see the size of her diamond!" or "When I get married, I want a really big stone!" In a nation where we have always been spoon-fed that "bigger is better," nowhere could this be more of a contradiction than in the case of a diamond. Carat weight is listed as the last of the "four Cs of diamond grading" for a reason: it is indeed the least important. It certainly is true that a larger diamond will always be more valuable than a smaller one of similar quality, but it is also true that a smaller but higher-quality diamond will in most instances be worth more than a larger gem of inferior grade. So first find the desired color, clarity, and cut among assorted diamonds; only then should you consider one with greater weight than others of similar quality.

The weight of a diamond, like that of most other gems, is expressed in carats or points. Before discussing weight versus size, let's get the meaning of the word *points* out of the way. This is easily understood even by the mathematically challenged, like me: one carat equals one hundred points, so one-half carat is fifty points, one-quarter carat is twenty-five points, and so on. Knowing the points of individual stones can be important in determining the value of a purchase you are considering, particularly in the case of clusters, bands, and bracelets containing many stones.

The cost to the dealer, and therefore to the consumer, is generally more for stones that are larger in size than for smaller stones of equal quality. Therefore, assuming the stones are of equal or very similar clarity, cut, and color, a diamond of greater weight (carats or points) will always be worth more than a larger number of diamonds with the same total weight (and therefore less weight for each of those stones). This is why a one-carat diamond solitaire can command thousands of dollars at retail, while a one-carat cluster ring of similar quality will fetch significantly less. As a rule of thumb, you can assume that cluster rings of smaller-weight stones will not be as costly as those consisting of heavier-weight stones of equal or similar quality.

Since we now know that one carat equals one hundred points, we can determine the individual weight of each stone in a cluster ring or similar piece of jewelry by counting the number of stones and then dividing that number into the total carat weight of the piece. For example, in a one-carat cluster ring composed of twenty-five equal stones, each stone is four points in weight; a half-carat band ring made up of five equal stones is set with ten-point diamonds. Of course, this is based on the assumption that all of the stones are the same. Still, in general this should make it easier for the casual consumer to compare the relative value of one piece of jewelry with many stones against another.

Since all this applies not only to diamonds but to most colored gems, synthetics, and even simulants as well, you will constantly find it useful when adding to your collection. Because this is so vital, let's double-check our knowledge before leaving this section. Consider the following example: While shopping for a one-carat diamond band ring, you narrow your choices to three. The first one contains ten stones; the second, twenty-five stones; and the third, fifty stones. If each of the three rings is selling for $500 and all other factors (such as gold content and fineness, similarity of setting, and quality) are equal, which one would be the best buy? If you choose the one containing ten stones, go to the head of the class! You have the concept down to a science.

Colored Diamonds: Natural or Enhanced?

Naturally colored diamonds are rare, and they fetch a very high price in the retail marketplace. Some brightly colored specimens can command tens of thousands, hundreds of thousands, or even millions of dollars. In fact, some of them are virtually priceless.

The diamond gets its color in much the same way that a colored gem does—through impurities that form within the crystal. Diamonds can also gain color through artificial exposure to heat, but these are of significantly lower value than naturally colored diamonds. The difference between a naturally colored diamond and one that has been enhanced may not always be easy to determine with the naked eye, and unscrupulous dealers can prey on unsuspecting consumers. Because of the additional costs incurred, even artificially enhanced diamonds can command a higher price than white ones of similar quality; however, the prices should be in the general vicinity of one another. For example, if you come across a one-carat white-diamond cluster ring that is selling for $500, and you are comparing it to a colored-diamond ring of similar quality and proportion priced at $650, it is a pretty safe bet that what you are looking at has been enhanced through a heat or pressure process of one sort or another.

An enhanced diamond must carry some kind of disclaimer that tells the consumer what he or she is looking at. One example of such a disclaimer would be the inclusion of the word *colored* in the description of the stone. Since this is not required of a naturally colored stone, an enhanced blue diamond should be clearly identified as "colored" or "enhanced," while one of natural color would simply be referred to as a "blue diamond." In addition, a gemologist or experienced jeweler can identify an enhanced diamond from one of natural color by performing a few simple tests. If you are in doubt, it is well worth having your purchase appraised. The appraisal may cost you a few bucks, but the peace of mind that follows will be priceless.

Please understand that there is absolutely nothing wrong with purchasing an enhanced colored diamond, as long as it is clear what you are getting and the price of the piece is realistic in terms of the marketplace. Since it is virtually impossible for the everyday gem shopper to detect a colored diamond that has been enhanced, you should stick with reputable sources if you decide to purchase one. Consider the years of experience and the place in the retail community that the particular firm enjoys, and shop accordingly. Again, if you are at all uncomfortable with your purchase, seek confirmation of the stone from an independent outside source. It is also critical here that the refund and exchange policies of the dealer be fully understood and clearly represented in writing.

Of course, common sense helps a lot too. Bear in mind that the cost of naturally colored diamonds is usually prohibitive for the average consumer. Rare finds, like a recent one of an orange diamond of five-plus carats, sometimes command millions of dollars on the market. Therefore, if the asking price for the piece under consideration seems to be in any way out of line, be suspicious. It can't be stated too often: if a deal seems too good to be true, it usually is.

If you are having difficulty parting with a relatively large chunk of cash for an enhanced gemstone, it may be easier to justify if you keep in mind that in many instances that color was already inherent in the stone, and the enhancement process simply brought it out. As further evidence of the value of such stones, remember that in most cases, only 25 percent (or less) of the stones picked for the treatment process actually end up as colored diamonds. Don't look at these advances in technology with a negative attitude, for they have put dazzling diamonds in a multitude of colors within the financial reach of the average jewelry lover.

Let the Buyer Beware: Impostors Abound

Until very recently, the diamond was imitated but never duplicated by humans. Advances in technology and research, however, have now produced the world's first synthetic diamond. Although this process has been hailed as a revolutionary breakthrough in the field of laboratory gem development, consumers have been slow to accept it because of its relatively high cost. But as top-quality diamonds continue to advance in price, the lab-created diamond may yet become a major player in the marketplace. Other competing stones can still be produced for a fraction of the cost of the synthetic, and finding ways to place this type of stone within the average consumer's price range is key to its future success. As of this writing, the market offers no other alternatives that can be classified as synthetics—that is, stones of approximately the same physical, chemical, and optical properties as the diamond. Because of this, the market for simulants, or imitations that merely look something like the real thing, continues to be strong.

One of the earliest diamond simulants was the paste diamond, which first appeared during the seventeenth century. Paste diamonds are composed of fused silica, soda ash, lead oxide, and other chemicals, including boron, a brightening agent used in gem treatments to this very day. Although the lead content of the paste diamond makes it bright in outward appearance, in reality it is a relatively soft material that does not wear well. The popularity of the paste diamond continues to wane with the availability of fine-quality cubic zirconia.

It should come as no surprise that the most popular diamond simulant today is the cubic zirconia, an impostor made by heating a mixture of yttrium oxide and zirconium oxide. These stones can be of remarkable clarity, and if cut properly and placed in diamond-like settings they can fool even the most experienced eye. However, because cubic zirconia is a simulant rather than a synthetic, a jeweler or gemologist can easily detect the difference within minutes of starting a close examination. Its vital properties, including its hardness, specific gravity, and refractive index, all differ from those of the diamond. One way anyone can detect a cubic zirconia is simply to breathe on it. The diamond, an excellent conductor of heat, will clear up quickly, while the imitation will not.

Aside from the cubic zirconia, other forms of diamond simulant are frequently encountered at retail. A stone of yttrium aluminum garnet (usually called YAG) was once popular but has nearly faded from view, thanks to the dominance of the cubic zirconia. Rhinestones, still quite popular today, can be made of either glass gems or quartz crystal. The brilliance of a rhinestone is improved by placing a metal foil on the bottom facets of a closed setting. All of these so-called fabulous fakes differ in composition, but they share one thing in common: they can easily be identified by a skilled jeweler or gemologist. A high-quality simulant can indeed fool the naked eye, but its crystalline structure will easily expose even the best diamond impostor.

The Future Outlook for the Diamond Supply

To properly evaluate the future outlook for diamonds, we must look at clear, white diamonds and naturally colored diamonds as two separate entities. The long-range availability of the standard, gem-grade white diamond is well established, while the future of the naturally colored diamond is inconsistent, at best.

The white-diamond market operates in a carefully controlled environment, which keeps this most precious gem within proper balance and availability. Prices stay relatively consistent, and overall the outlook for supply remains optimistic. New sources in unlikely places such as northwestern Canada continue to fortify the diamond market. Speaking of unlikely places, the first commercial diamond mine in the United States opened in June 1996 at Kelsey Lake, Colorado. Located north of Fort Collins near the Wyoming border, the mine is small by world diamond-mining standards; however, the quality of the initial finds is said to be excellent. This site is expected to provide lovely diamonds perhaps until the end of the next decade.

Although the supply of white diamonds seems at this point to be almost endless, the important naturally colored ones face an entirely different future. Even at the current rate of production, naturally colored diamonds account for just the slightest percentage of the total, compared with production of the white or colorless variety. This obviously means that naturally colored diamonds will continue to command higher prices in the marketplace with the passing of time, eventually placing them well beyond the reach of the average consumer. If you are among those fortunate enough to have a naturally colored diamond in your collection, consider it a natural treasure and take all precautions necessary to keep this rare find alive for generations to come by preserving it and passing it on as a family heirloom.

Chapter 9

The Most Popular Colored Gems

As previously noted, the imaginary boundary that once separated the four precious gems—diamonds, emeralds, rubies, and sapphires—from the rest has pretty much disappeared. All but the most stoic have really embraced the modern-day concept of one general group. Although it did away with the debate that raged over which should be precious and which should not, it has given rise to yet another controversy: establishing a pecking order of the "gem society." In order to keep the debate on this matter to a minimum, I have decided to list all the most popular mineral groups in alphabetical order. (Thus, for example, all beryls—including emeralds—are described first, and rubies and sapphires are grouped with other corundums.) Unfortunately, this puts zoisite (the mineral group that includes tanzanite) dead last on the list. Sometimes you just can't win.

Beryl

Without a doubt, the beryl gem family is a most important mineral group. Its family of gems includes the aquamarine, emerald, goshenite, heliodor, and morganite, as well as a little-known red variety and another green beryl that is not quite emerald material. Colors range from soft pastel to vivid, and all are excellent additions to the jewelry wardrobe of the gem collector.

Aquamarine

If asked to guess which member of the beryl family is considered the most endangered, most consumers would probably say the emerald. Certainly, top-quality emeralds of any considerable size are not seen frequently at retail, yet the future of the aquamarine is even more uncertain. At one time the aquamarine was considered to be abundant, and market prices reflected this. Today it is not uncommon for a good-quality aquamarine in a larger carat weight to command hundreds of dollars or more per carat.

The world's taste for aquamarines has changed with the times. The sea green variety was for a long time the most highly prized; in fact, its name literally means "seawater." The earliest sailors would take aquamarines along and toss them in the water to satisfy the often angry god of the seas, Poseidon. At

today's prices, I doubt anyone would throw his or her prized aquamarines into the sea, no matter how mad Poseidon was!

Today's consumer looks more for varieties displaying shades of blue that run the spectrum from sky blue to ocean blue. It is worth noting that a sea green aquamarine, with undertones of yellow, can be heated to produce a strikingly beautiful shade of blue. This practice is common with virtually all aquamarines found at retail today, and therefore does little to affect its market value. Aquamarine has also been seen in a rare cat's-eye state; this variety is always cut *en cabochon*. When shopping for this lovely blue beryl at retail, be forewarned that high-quality, heat-treated blue topaz has sometimes been passed off to unwary consumers as aquamarine.

Aquamarine is accepted as the modern-day birthstone for March, but those born in October also recognize it as their birthstone. If you are born under the sign of Aquarius or Scorpio, you should know that the aquamarine is one of your astral stones as well. Aquamarine comes from many different places in the world, but some of the most desired ones originate in Brazil, in the mining region of Minas Gerais. Other important sources include parts of the former Soviet Union, China, India, Pakistan, and Nigeria.

Emerald

Unquestionably the most popular member of the beryl group is the lovely green variety the world knows as the emerald. Throughout history, the emerald has played an important role in society, and it is a stone rich in folklore and tradition, embraced by many cultures. The emerald is considered not only the traditional birthstone for May but also a nontraditional birthstone for January, June, August, and September. Not wanting to be outdone, astrologers have adopted the emerald as an astral stone for the signs of Taurus, Gemini, and Cancer, and they link it to Mercury, Venus, and the moon. The emerald is considered appropriate for couples celebrating a twentieth, thirty-fifth, or fifty-fifth anniversary.

The legendary "mines of Cleopatra" in Egypt are believed to be among the earliest emerald mines on Earth, and there is evidence of mining activity in this region as far back as 2000 B.C. (centuries before the time of Cleopatra). These mines accounted for most of the emeralds supplied to Western civilization for millennia, and they continued to prosper well into the eighteenth century. Those who labored in the unbearably hot and dry conditions there were in constant danger of death from heatstroke and dehydration. Finally, around 1750, the mines were believed to be exhausted and were abandoned. They remained dormant until the early twentieth century, when an effort was made to reestablish them as a source of supply. This effort failed miserably, as the mines produced stones that were generally regarded as less than gem quality, and the area was finally abandoned for good in the early 1920s.

Today there are a number of sources of the emerald, led by Colombia and also including Brazil, India, Australia, South Africa, Pakistan, and Zimbabwe. Colombia (in particular, the mines of Muzo and Chivor) is credited with producing the highest-quality emeralds in the world, but this is not to say that other sources are unimportant; all are capable of producing emeralds that rival those from Colombia in quality.

Emeralds grow naturally within the Earth, formed under intense heat and pressure, and are mined by conventional shaft-mining techniques. They also can be found in alluvial deposits by natives who search the sometimes treacherous waters for their precious finds.

Once, the only gem to be classified as an emerald was the green beryl colored by chromium, but all that changed about thirty years ago when an important discovery was unearthed in Brazil. These beryls were the first of their kind ever found to be colored by vanadium rather than the traditional element of chromium, and this history-making event sent ripples through the gem world. An important find of Brazilian emeralds was unearthed in 1995 at Nova Era, which is in the gem-rich state of Minas Gerais. This find was welcome news for Brazilian gem traders, who had become concerned in recent years with dwindling supplies elsewhere in the country. Although some may think of amethyst when they think of the gems of Brazil, the emerald is actually that country's number one gem of export. Another intriguing development in Brazil took place in late 1997, when a farmer discovered emeralds while tending his fields. Much of the material reportedly contains inclusions, and this has given rise to talk of a cat's-eye emerald.

Today the lab-created emerald so dominates the retail scene that I think it worthy of a brief discussion. Born basically to take the place of the highest-quality prized emeralds in the marketplace, these synthetics offer the consumer an affordable and beautiful alternative to those rare and costly gems. Lab-grown emeralds are made by exposing thin slices of natural beryl to a chemical mixture similar to that of a natural emerald. These materials, together with a mixture of pure filtered water and acids, are then heated under intense pressure in a vessel made of top-grade stainless steel, then cooled in a controlled environment until the crystal forms. This process is repeated over and over until the crystal reaches a thickness that is substantial enough to be cut and faceted into gem form. Other man-made gems such as rubies and sapphires (even the cubic zirconia) have evolved from a process using this method as its starting point. Naturally, the compositions of the mixture vary from stone to stone, but the heat-and-pressure method remains the same for most created gems. It has worked well enough for Mother Nature, so why not for humans?

No matter how precisely people can duplicate the natural emerald, however, there still seems to be something missing—an intangible feeling that owning a treasure grown within the Earth seems to provide. Unfortunately, while top-quality natural emeralds are still available, their cost has become so prohibitive that many dealers will sell them only on a special-order basis. Naturally, the price of your must-have treasure will vary from stone to stone, but one thing is absolutely certain: if it's a high-quality clear solitaire you are after, be prepared to spend major bucks. As a guideline, if you dabble in real estate and own, for example, the island of Bermuda, it's probably safe for you to go ahead and place your order.

Since not many consumers own an island, this puts most of us in a quandary. Do we go with a lesser-quality natural emerald, content in knowing that it may not be a $10,000 gem but is still direct from nature; or do we go for the lab-created gem that duplicates the top-quality natural one? This is one question my book can't answer, for the choice depends on financial wealth and personal preference. Set your fiscal priorities, or just follow your heart.

Goshenite

Also known as the white beryl, goshenite is the hard-to-find colorless variety of the beryl group. The demand for goshenite is pretty small, and as it turns out this is just as well, because it is unlikely that sources would be able to satisfy a large demand. When found at retail, they are nearly always seen in small sizes with a weight of five points or less (and remember, there are one hundred points in a carat), generally serving as accent stones for their colorful relatives, including the emerald and the aquamarine. Don't bother holding out for a five-carat oval, as large solitaires are seldom seen. On occasion, clusters or bands consisting entirely of goshenite can be found at retail, and these also would make excellent collectibles for any gem lover. Goshenite gets its name from the area in which it was first discovered, the town of Goshen, Massachusetts. I'm sorry to report that the original finds were quickly exhausted, leaving Brazil and parts of the former Soviet Union as its major sources. Amateur prospectors may take note that small finds occasionally turn up in the states of California and Maine. Some astrologers link goshenite to the planet Venus.

Heliodor

Known in the jewelry trade as the golden beryl, the heliodor is often confused with other gems of similar color. When found in its deep golden shades, one is reminded of the citrine found in Madeira, Spain, but this heliodor is most often found in parts of the former Soviet Union. Heliodor is also seen in a light yellow

shade, and this color is generally associated with the gem-rich country of Brazil. Although iron is its principal coloring agent, the heliodor also contains trace elements of cesium oxide. The heliodor was once considered an aid to kidney problems. In earlier times, a dream about a heliodor portended the arrival of great happiness. The heliodor is often recognized as a nontraditional birthstone for November, and astrologers link it to the planet Jupiter. Sometimes found in conjunction with the aquamarine, the heliodor is a relatively inexpensive gem that should be a part of your personal wardrobe.

Morganite

Sometimes confused with the pink tourmaline or the kunzite, the morganite is a variety of beryl ranging from pastel pink to soft peach. In fact, the first stones of its kind were discovered in Southern California by an expedition mining the region for pink tourmaline. Named for the wealthy gem fancier J. Pierpont Morgan, the morganite first appeared in the early twentieth century. Even though it isn't exactly a high-profile gem, those born in October sometimes claim it as their birthstone.

The morganite gets its color from manganese, while its higher specific gravity is due to the presence of cesium and lithium, two rare elements in nature. Like the aquamarine, the morganite is dichroic in nature (exhibiting two different colors when seen from different angles), and occasionally aquamarine and morganite are even found growing together in the same resident crystal, producing a stunning effect. These bicolored gems from the aquamarine-rich lands of Minas Gerais, Brazil, are very rare, available only to wealthy investors through personal contacts.

Because of its lack of recognition, the morganite is not being heavily mined anywhere in the world. The earliest finds in Southern California proved to be unreliable, and the only other source of any significance is the gem-laden island of Madagascar, off the southeastern coast of Africa. Like the goshenite and the heliodor, the seldom-seen morganite would be another excellent find for the amateur gem collector.

Other Beryls

As indicated earlier, there are two little-known varieties of beryl that few consumers will ever see, but they are worth a quick mention.

A rare red variety of beryl originates in the rugged mountains of Utah. Like the emerald, it lacks the dichroic nature of the aquamarine and the morganite. Red beryl gets its vivid color from manganese that is found resident in the crystal. Formerly known as bixbite, this rare and beautiful gem is seldom, if ever, encountered at retail. Top-quality red beryl is highly prized, and some consider it to be rarer than even the finest of emeralds. Since it has been found in just this one solitary location, scientists believe that only the climate there is suitable for its crystallization process. Recent reports coming out of the region indicate that a new deposit has been discovered. How much rough is available and how it will affect the market are two concerns yet to be addressed.

Most consumers are surprised to learn that there is another variety of green beryl that lacks sufficient green color to be classified as an emerald. This variety, which can be found wherever deposits of emeralds occur, is considered less than gem-grade material.

Besides these two varieties, a dark blue beryl was uncovered in Minas Gerais, Brazil, in the early twentieth century. It proved to be inconsistent in supply, and scientific tests eventually confirmed that this blue beryl faded in the sunlight, losing its dark blue color in a relatively short period. From time to time, other colors of beryl also surface, although they are usually gobbled up by gem fanatics long before they can reach conventional retail outlets. Some colors observed include brown, orange, and lilac. Unless you have some serious connections with a gem dealer, you are not likely to come across any of these rare, highly sought-after finds anytime soon.

Chalcedony

Comprising storied gems that have been cited since history began, the chalcedony (pronounced "kal-*sed*-on-ee") group is actually a form of quartz. It has a large immediate family, consisting of many varieties of agate, as well as bloodstone (sometimes known as heliotrope), blue chalcedony, carnelian (also called cornelian), chrysoprase, jasper, onyx, sard, sardonyx, and even petrified wood. Because of its link to the mineral quartz, chalcedony also has many distant relatives. These include amethyst, citrine, tiger's-eye, and cat's-eye quartz. In a strange and indirect way, the chalcedonies are even related to the precious opal. All of these are quartz mixtures with a variety of structures. The members of the chalcedony group are mostly translucent to opaque in appearance, and they are usually seen in bead form, as drops, cut *en cabochon*, or with a flat cut. Some astrologers link chalcedony in general with the planets Jupiter and Saturn. Let's take a brief look at some of the many members of this group.

Agate

Agate occurs in a wide assortment of colors, but the ones most commonly seen are blue and green. Most agates are banded, though there are exceptions. Moss agate is a multishaded green variety that resembles moss from afar. Since it isn't banded, some gem authorities don't consider it a true agate. Moss agate is sometimes found with bands of brown agate; this variety is known as mocha stone. Two of the many other types of agates are the iridescent fire agate and the fortification agate, which has sharply angled bands. Early Egyptians carved agate into receptacles and amulets, as well as other articles of adornment. In addition, agate has been used for thousands of years in the creation of the carved cameo. Today's technology allows intricate figures such as angels, animals, and Victorian women to be laser-inscribed into ovals of beautiful blue and sea green agate.

Agate is a nontraditional birthstone for May and June. Those born under the sign of Gemini, Taurus, Virgo, Libra, or Capricorn all claim agate in one form or another as their astral gem. Astrologers link agate to the planet Mercury. Couples celebrating their twelfth or fourteenth anniversary consider agate their anniversary gem.

Agate is found in deposits in many parts of the world. The Cumberland Plateau, which spans Alabama and Tennessee, is one of the best sites for agate collectors in the United States. The area abounds in many varieties of minerals, and the chalcedonies are well represented here. Impressive mountains, natural waterfalls, and the beautiful lakes that were formed as a result of government control of the Tennessee River provide a feast for the senses to the prospective rock hound. A number of states list the agate in various forms and colors as their state stone; these include Kentucky, Louisiana, Nebraska, and South Dakota. Agate is also the international gemstone of the country of Denmark. According to folklore, a dream about agate means a long journey is just ahead.

Bloodstone

Ancient Christianity held that the blood of Christ dripped from the cross onto a dark green stone that lay at his feet. This was believed to be the origin of bloodstone, sometimes called heliotrope—but not to be confused with hematite (see chapter 10), which in some places is also called bloodstone! A gem rich in folklore, bloodstone at one time was believed to be a cure for the common nosebleed. Even today, desert-dwelling tribes wear bloodstone around their necks to protect them from the poisonous bite of the scorpion and the rattlesnake. A dream about bloodstone was once thought to bring about the onset of distressful news. Personally, the name alone scares me.

This hard-to-come-by member of the chalcedony family is usually dark green with specks of red, which are caused by the presence of iron in the mineral. (There is another, very similar variety of chalcedony called plasma, which has spots that tend to be yellowish.) Bloodstone is most likely to be found in India; other sources include Brazil and Australia. It is the traditional birthstone for March, and some people connect it with December. Those born under the sign of Pisces, Aries, or Scorpio all claim bloodstone as their astral stone, and astrologers logically link it to the red planet Mars, named after the Greek god of war. Because of its stark appearance, much bloodstone is set into bezels and fashioned into men's rings. For whatever obscure reason, some men do not feel it is appropriate for a woman to wear bloodstone. To date, there are no man-made equivalents of bloodstone.

Blue Chalcedony

Blue chalcedony is one of the most popular forms of its mineral group today. This material may be found in Turkey and Namibia, but most experts agree that the finest blue chalcedony originates in the United States. Until recently, blue chalcedony was pretty much limited to the Pacific Northwest. California, Oregon, and the state of Washington all produce lovely specimens, although none of these deposits are considered long-term. However, a new discovery, named the Mount Airy blue, in a remote section of central Nevada is said to be not only of fine quality but also of significant size. Besides the translucent blue color, some gems from Mount Airy display a cat's-eye effect or a mysterious haze, from silver-blue to pink, that is known as adularescence. This phenomenon is more commonly associated with moonstone, a member of the orthoclase family of the feldspar mineral group.

Blue chalcedony is found in large rocks known as thunder eggs, which are often associated with beds of ash. Although they are generally less than one foot in diameter, these rocks can be as large as three feet or as little as two inches across. Individual pieces of rough may exceed five pounds, though most average less than a pound.

Carnelian

In carnelian, which is also called cornelian, the density of color depends on the amount of iron in the mineral. Found primarily in India, as well as various sites in South America, the most favorable pieces are those of a hue from deep red to red-orange. While it is usually found in nature with colored bands, some specimens gain uniformity of color as a result of either exposure to the sun or a treatment process involving heat and pressure. Some traditions consider carnelian the birthstone for July, but the ever popular ruby is more widely accepted; others regard carnelian as a birthstone for May and August. It has a long and storied past, and as a result those born under the sign of Aries, Taurus, or Virgo all claim carnelian as their astral stone. Astrologers link it to the planet Mars. Carnelian was once considered strictly the property of the noble class; in fact, those of high social status were often buried with the gem. It was sometimes also used in the carving of cameos, although this is much less common today.

Chrysoprase

Gem experts consider chrysoprase to be the most valuable member of the chalcedony mineral group. The stone gets its apple green color from the presence of nickel in the crystal. (An extremely similar but more somber-looking variety of green, called prase, is very rare.) Consumers should note that if the stone is exposed to intense sunlight, the color may begin to fade. In fact, worthless chrysoprase of poor color is sometimes passed off to unsuspecting consumers as fine-quality jadeite. It comes from many places, including Australia, Brazil, India, and various locations in Africa; in the United States, California is considered the most important source. Chrysoprase is a nontraditional birthstone for May and December, and astrologers link it to Mercury and Venus. Unfortunately, it is seldom encountered at conventional retail outlets; nevertheless, low demand has kept its price relatively affordable. The everyday gem buff should make chrysoprase a part of his or her gem wardrobe. Just don't leave it on the dashboard in the blazing summer sun!

Jasper

Jasper is a gem seen in a variety of colors that may occur separately or together. For example, there is the almost mystical stone called the picture jasper, which usually comes in multiple shades of brown. Although most experts consider it part of the chalcedony mineral group, this most unusual gem is actually a mixture of quartz, opal, and chalcedony, with impurities of clay and iron. Jasper, like most chalcedonies, is found worldwide, with deposits particularly in India, France, Germany, and parts of the former Soviet Union. California is considered the most important U.S. source of jasper, though the hot, dry deserts of Nevada also yield some pretty specimens. Most of this material is considered too small for gem use, so jasper from the desert usually finds its way into the market as a tumbled stone, either alone or in artifacts. Those born in March should know that jasper is one of your birthstones. Astrologers hold the gem in high regard. It is attributed to the signs of Scorpio and Sagittarius and is linked to the fixed star of Libra; because of its multicolored bands, jasper is also often considered a stone of the planet Saturn, but it is sometimes linked to Mercury and Venus as well. Jasper is seldom seen in jewelry and the average consumer does not often consider purchasing it, but it remains affordable to most and is an excellent choice when filling in your unusual gem collection.

Onyx

The rich black gem that the entire jewelry world knows as onyx usually isn't onyx at all, but rather a form of agate that gets its uniformity of color through a relatively simple method of treatment. During this process, agate (usually gray) is soaked in a dense solution of sugar and water—sometimes honey is used—for a period of approximately one month, after which sulfuric acid is introduced into the mixture. The acid produces carbon, which turns the agate black. Since agate and onyx share similar properties and values, this material is pretty much accepted worldwide, and consumers seldom give it a second thought. Another stone, with white and brown bands, that is commonly known as Mexican onyx isn't onyx either, but rather a mineral known as calcite. This Mexican onyx is seen seldom, if ever, in the jewelry world, appearing instead as carvings and artifacts. There is absolutely nothing wrong with purchasing this material; just be aware of what it is you are actually buying.

By contrast, natural onyx is most often seen as a black-and-white-banded variety of gemstone that has been used in the carving of artifacts for centuries. Many of these ancient pieces, dating back to the days of the Romans and Egyptians, are still with us today. South America is generally considered the world leader in onyx production, but onyx is a plentiful gem found worldwide. Various traditions list onyx as the birthstone for February and July. Those born under the sign of Cancer, Leo, or Capricorn often claim the onyx to be their astral gem.

Petrified Wood

Petrified wood has been around practically since time began. It actually starts its life as organic matter. Under certain conditions, as it decomposes the organic materials are replaced by (*not* transformed into) agate, bit by bit. Jasper, quartz, and even opal also can be found in wood that has fossilized.

Petrified wood is most often brown, but it may also be gray or green in appearance. The most famous site for petrified wood is Petrified Forest National Park, near the town of Holbrook, Arizona. It can also be found in Nevada. Outside the United States, sources of petrified wood include Argentina, Egypt, and the Czech Republic.

Petrified wood is sometimes also found buried within sequoia, oak, and cypress trees as they decay beneath the surface of the Earth. About the only other thing I can contribute to this section is the fact that my wife recently gave me a large, heavy pair of petrified-wood bookends for Christmas. I would like to go on record here and now in declaring this the heaviest present I have ever received. These two six-by-four-inch bookends weigh just slightly more than a 1963 Dodge.

Sard and Sardonyx

Sard is the reddish brown variety of chalcedony, usually seen in bead form when encountered at retail. Sardonyx, on the other hand, is easily distinguished from sard by its multicolored bands of reddish brown and white. Like sard, it is commonly found in bead form, and both varieties are seen more often as art objects than as jewelry. Pakistan is generally considered the world's leading source of both sard and sardonyx; however, both forms are found worldwide, usually among other chalcedony deposits. Sard is usually found in colors of red to brown, while sardonyx combines the color of sard with the white bands of onyx, which produces a stunning effect. For this reason, sardonyx is sometimes called "the fancy wallpaper of nature." Those born under the sign of Aries or Scorpio sometimes claim sard as their astral stone. Sardonyx is the traditional birthstone for August and a nontraditional birthstone for July and September. Some astrology buffs link sardonyx to the signs of Virgo and Scorpio. It was once considered a gift to the Earth by the planet Saturn, no doubt because of its multicolored bands. Others link sardonyx to the planet Mars. In earlier times, craftspeople used sard in meticulous inlaid artifacts. Sardonyx was first used in the creation of the cameo, though this is rarely seen today.

Chrysoberyl

The casual collector is for the most part unfamiliar with the chrysoberyl mineral group. Even those who recognize (and perhaps even own) natural alexandrite are usually unaware of its other family members. In fact, that isn't even considered the most sought-after gem in this group (which, incidentally, astrologers link to the sun). Now that I have aroused your curiosity, let's take a closer look at this lovely group of gems.

Alexandrite

Certainly no one would dispute that the alexandrite is the most famous variety of this beautiful and rare mineral group. This spectacular gem is colored by chromium occurring in combination with iron. High-quality alexandrite will appear a vivid green color when viewed in daylight and a bright red hue in incandescent lighting. Natural alexandrite is thus an excellent example of what the gem world calls the color-change phenomenon.

Natural alexandrite is so rare and expensive that less costly alternatives produced in controlled laboratory environments now have taken hold in the marketplace. Consumers should take note that exposure to extreme heat, such as a jewelers torch or steam cleaner, can alter the color-change properties of the stone. So be careful when selecting a jeweler to size an alexandrite ring.

The gem was first discovered in Russia in the early part of the nineteenth century and subsequently named after Czar Alexander II. Many people still refer to the stone as "Russian alexandrite," but the Russian deposits produce very little of this rare gem today. The two most important sources of alexandrite in the world now are Brazil and Sri Lanka. Alexandrite is a nontraditional birthstone for June and August. Those born under the sign of Gemini lay claim to it as their astral gem. It is also used to celebrate a forty-fifth or fifty-fifth anniversary. To dream of an alexandrite was said to foretell the onset of hard times and financial difficulty. To me, if you're lucky enough to own one of these rare and costly beauties, you've got to have at least a few bucks hanging around somewhere. Why worry now?

Cat's-Eye Chrysoberyl

It may surprise you to know that the rarest member of the chrysoberyl mineral group is not natural alexandrite but a beautiful color-change variety that displays the unusual cat's-eye phenomenon, also called chatoyancy, resulting from very narrow inclusions lined up in one direction within the stone. The most highly prized form of cat's-eye chrysoberyl is a honey brown variety, but it comes in other colors as well, all of them considered rare. This most mysterious gem is an astral stone of Pisces and Capricorn, and astrologers link it to Ketu, the southern lunar node, where the sun and moon appear to cross paths in the celestial sphere (resulting in eclipses and, according to some who have studied the stars, various unexplained phenomena). Cat's-eye chrysoberyl is found primarily in Sri Lanka and Brazil; some material also surfaces in China from time to time.

Although other minerals display this cat's-eye effect, the chrysoberyl is the variety that many gem purists consider the only true form of cat's-eye in the world, and only it can be referred to simply as a cat's-eye. Other minerals found to possess this effect must carry their common mineral name appended like some gemstone scarlet letter, branded for life: for example, the cat's-eye quartz—a striking gem, yet forever doomed to secondary status.

The future of both the alexandrite and the cat's-eye chrysoberyl can best be described as uncertain. No important new sources have developed, and current deposits are sporadic at best. Large alexandrite stones are found these days only in museums or exclusive private collections, and the even more elusive cat's-eye chrysoberyl is next to impossible to find in any size at any retail venue. Fine-quality man-made copies first appeared on the market in the mid-1970s; considering the cost of their natural counterparts, they represent the only viable option for the average gem collector. If you happen upon either of these two gems, you'd better save your lunch money; trust me, you're going to need it.

Yellow Chrysoberyl

Still another member of the chrysoberyl group, this gem may actually be either yellow or yellow-green in color. Although it is more common than the other two varieties of its family, the yellow chrysoberyl is seldom encountered at retail. Found today primarily at gem and mineral shows and rock shops, the yellow-green variety may be easily confused with peridot. As if its color weren't cause enough for confusion, both the peridot and the yellow-green chrysoberyl at one time were known as chrysolite. The yellow chrysoberyl may be located anywhere the alexandrite is found, but its primary sources these days are considered to be Brazil and Sri Lanka.

The future of chrysoberyl seemingly has been in question forever, and it remains on the endangered list to this day. Recently, finds out of eastern India have been reported, giving some relief to a beleaguered marketplace. All varieties, including alexandrite, yellow chrysoberyl, and even cat's-eye, have been discovered, though some of the material (particularly the cat's-eye) is said to be heavily included. As of this writing, the scope of this deposit is undetermined.

Corundum

The corundum mineral group is one of the most highly prized of all. Its three family members are the padparadscha, the ruby, and the sapphire. All corundums have a hardness of 9.00, which makes them the hardest colored gems in the world. Only the diamond, at a perfect 10, rates higher on the Mohs' scale. The corundums have been so popular for such a long period of time that the first lab-created gems date all the way back to the turn of the century.

Padparadscha

The padparadscha, one of the rarest gems on Earth, occurs in nature when trace elements of chromium and vanadium combine with iron. It is seldom seen at retail, and its strongest color is best described as orange, with hints of pink, yellow, and at times peach. Its color has been compared to that of a lotus flower, and padparadscha is sometimes called "the king of gems."

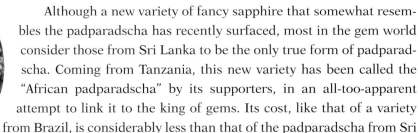

Although a new variety of fancy sapphire that somewhat resembles the padparadscha has recently surfaced, most in the gem world consider those from Sri Lanka to be the only true form of padparadscha. Coming from Tanzania, this new variety has been called the "African padparadscha" by its supporters, in an all-too-apparent attempt to link it to the king of gems. Its cost, like that of a variety from Brazil, is considerably less than that of the padparadscha from Sri Lanka, however, and it may become a viable option for consumers in the coming years. If its color is even close to that of the true padparadscha, it will be a wonderful find, and a must for your jewelry collection. Most gem experts refer to this gem simply as a fancy African sapphire.

I have not yet seen the African variety firsthand, but I have had the opportunity to hold (in my constantly shaking hands) the true padparadscha from Sri Lanka. An oval-cut gem just over one carat in weight, it had an established value estimated at approximately $15,000—unset! Since most of us don't carry that kind of pocket change around too often, the lab-created padparadscha has recently entered the marketplace. Affordable and, of course, readily available, this man-made form of corundum offers the amateur collector an opportunity to own a beautiful gem unlike any other. Unless you are a person of incredible means, I would recommend you add at least one of these lovely lab-created creations to your collection; in fact, I already have added it to mine. It is one of my wife's favorite gems.

The future of the true padparadscha remains in doubt, and someday the fancy African sapphire may be the only viable choice for those seeking a natural alternative to this beautiful, endangered species. Even though this rare variety of corundum is already well beyond being affordable for the average consumer, it will be a sad, dark day when the world's supply of padparadscha is exhausted.

Ruby

The ruby has long been a favorite of consumers worldwide. The ruby is technically red corundum; unlike other corundums, the red variety has never been tagged as a type of sapphire but has always had its own separate name. Rubies come in many shades of red, from the most sought-after variety of pure red, known as pigeon's blood, through mixtures that display undertones of blue, purple, orange, or brown. Top-quality natural solitaires of any considerable carat weight are already well beyond the means of the average consumer, leaving little alternative besides lab-created gems. Only you can decide whether to go for the looks of the finest synthetic or to sacrifice the sparkle for the romantic feeling that only a naturally mined ruby can create.

It is pretty much accepted among dealers that the highest-quality natural rubies originate in the gem-rich Mogok region of Myanmar, the country that is officially no longer called Burma. This gem is known to experts and consumers worldwide as the Burmese ruby. The earliest record of rubies from Burma dates back to the late 1500s. Though still the most important ruby-producing area of Myanmar, this region is not the country's only significant source. In fact, excellent specimens of smaller size are found today in the northeastern part of the country, in the area known as Möng Hsu (pronounced "Mahng Shoe"). The gems from this region first began to appear in 1992, and they are generally seen in sizes of three carats or less.

Initially, all rubies mined in Burma were considered the property of the government, and early accounts detailed incidents in which entire villages were massacred for smuggling these gems out of the country. Even though strict legislation remains in place, some Burmese rubies have made their way into neighboring Thailand by way of the Thai border town of Mae Sai. Just as the sapphire from Sri Lanka continues in the jewelry trade to carry the older name Ceylon, the ruby from Myanmar continues to be known as Burmese. Unquestionably, this is a result of its tremendous name recognition and reputation for quality. In fact, dealers also refer to top-quality gems from other areas as "Burmese grade," a practice that is generally accepted throughout the industry.

Supply of the Burmese ruby has been described as sporadic and inconsistent. In order to keep prices high, the Myanmar government carefully manages the marketplace, creating what are known as spot shortages in the industry. During these times of shortage, more-defiant gem smugglers literally risk life and limb to transport these stones across restricted borders, commanding inflated prices from customers waiting with open arms.

Myanmar may be the highest-quality source of rubies, but Burmese rubies do not completely dominate the market. Thailand also has been known to produce some top-quality specimens, and gems originating from Pakistan have often made the grade as Burmese. On the other hand, although the gems from Sri Lanka can be lovely, they are usually found as waterworn pebbles in alluvial deposits, which makes them inconsistent in color and limited in size. African nations provide large gems that sometimes carry an undertone of blue to purple. Budget-minded consumers in the market for a large natural solitaire can sometimes find a bargain here. In a 1996 development, a site formerly known as the John Saul Mine reopened in Kenya; how much of an effect it will have remains to be seen. The United States is not considered a serious source of gem-quality rubies; still, mines throughout the country offer novice prospectors a chance to try their luck. Don't go with the expectation of making millions, for these expeditions usually yield stones considered to be low-end, or less than, gem quality. Quality aside, the excitement that comes with the discovery of a natural gem is a feeling like no other. Try it sometime: become an amateur rock hound for a day!

The ruby is the traditional birthstone for July, though others consider it a birthstone for December. It is used to celebrate a couple's fifteenth and fortieth anniversaries. The ruby is considered the most powerful gem in the universe, and as such is associated with a variety of astral signs, including Cancer, Leo, Virgo, Scorpio, and Capricorn. It is often associated with the sun, though some link the ruby to the planet Mars. Personally, my beautiful wife, Linda, and I have always considered the ruby to be our spiritual link, since it is my traditional birthstone (July) and her traditional astral stone (Capricorn). That probably doesn't mean much to you, but it sure was enough to make me fond of the gem.

Rubies overall are in no immediate danger of extinction, but high-quality, clear stones of pigeon's-blood red are becoming scarcer and are already quite costly by most standards. Unless there is a dramatic change, this situation will probably only get worse over time. No other sources of rubies as of yet have consistently measured up to the natural, crystal-clear variety of those from Myanmar. However, a recent deposit uncovered in Kashmir, India, has given some cause for at least cautious celebration. Since most gem historians consider the blue sapphires from Kashmir to be the highest-quality ones ever found, early speculation about the rubies is encouraging. Most of the initial goods from there have been three carats or less, but some experts have already labeled them as Burmese grade. Only time will tell.

Star Ruby

Like many other gems, the ruby is also sometimes found to contain a six-rayed star, a phenomenon known in the gem world as an asterism. This effect is caused by intersecting needle-like inclusions called rutiles. To maximize the appearance of the star when offered at retail, experts always cut this gem *en cabochon* and polish it to a high finish. Although collectors seek out star rubies, this is due more to their rarity than to the quality of the stones. In reality, experts rarely, if ever, consider any gem that exhibits an asterism to be a top-quality one. Still, if you come across a ruby (or any other gem for that matter) that displays this phenomenon, by all means add it to your collection if it fits your budget.

Because of its red color and starlike effect, the star ruby is regarded by some astrologers as a stone of the sun. For centuries, early cultures of the Orient considered it a sign of good luck and fortune. In ancient times, wise men were thought to carry star rubies with them wherever they traveled, as they believed the star to be the source of great wisdom. The six-pointed star is said to represent the virtues of faith, hope, and charity.

Sapphire

It's time for a little word-association game. Close your eyes and tell me what you see when I say the word "sapphire." If you are like most people, visions of a lovely blue gem come immediately to mind. Certainly no one can argue that the most popular sapphire of all is blue, yet the sapphire comes in a wide variety of other shades as well. All colors of sapphire other than blue are commonly referred to in the gem world as "fancy sapphires," and they are discussed in detail separately.

The blue sapphire is one of the most popular colored gems in the world, consistently among the top two or three in total sales, along with the emerald and ruby. There are actually many different shades of blue, and as with the fancies, these various colors sometimes indicate their country of origin. Let's examine some of them; perhaps it will help you to identify your own blue sapphire.

The sapphire has had more than its share of folklore and tradition. Some ancient societies believed that the Earth rested upon a sapphire, which they theorized made the sky blue. Legend has it that if a sapphire was worn by a person of evil intention, it would get angry and refuse to shine. Sapphires dipped in cool water were believed able to cure sight disorders. The sapphire is the traditional birthstone for September and a nontraditional birthstone for January, June, and November. It is also considered an astral stone for the signs of Taurus, Virgo, and Libra, and astrologers link it to the planet Saturn. Couples use sapphires to celebrate a fifth or forty-fifth anniversary.

As previously noted, most gem historians agree that the finest-quality blue sapphires ever mined originated in Kashmir, India. These deposits, once considered the best source of sapphires in the world, have been depleted for a great many years. Today, true Kashmir sapphires are found only on the secondary market, and in many instances they have become family heirlooms. The name Kashmir blue is still used by many dealers when describing the appearance of a top-quality, royal blue sapphire, regardless of the source. As in the case of the Burmese ruby, this is pretty much accepted by experts worldwide. Although at first it may seem a bit misleading, this actually does help to separate the top-quality sapphires from similar ones of lesser grade. This terminology seldom causes confusion in the sapphire trade, as most dealers already know that the original source of Kashmir sapphires went dry long ago.

By now, chances are you have already decided that nothing but the best will do, and that one way or another you are going to acquire a Kashmir sapphire. This may not be as hard to do as it may seem—particularly if you own Brazil. If not, you can easily find a variety of synthetics that will fool your friends into thinking you do.

Many people believe that the premium blue sapphire on the market today is the Ceylon sapphire, a lovely cornflower blue corundum. This stone is lighter in color than the majority of others within the group, though beautiful blue specimens that rival the Ceylon in color can be found almost anywhere fine-quality sapphires originate. This lovely specimen continues to be known as the Ceylon sapphire even though the country of origin is now known as Sri Lanka. As with Burmese-grade rubies, Ceylon-grade sapphires continue to enjoy the highest name recognition and market acceptance.

Some of my favorite sapphires, including the two rings I have given to my wife, originated in Kanchanaburi, Thailand. A rich and fertile land of many resources, Kanchanaburi has played a significant role in the production of sapphires for many years. Sadly, nothing lasts forever, and this once bustling area can no longer be counted on as a consistent source of gem-quality rough. Although the world-famous Chanthaburi mine continues to produce lovely blue specimens, new finds in other lands have caught the attention of investors and miners alike. It is still considered the gem capital of the world, but Thailand's golden era may indeed be flickering out.

The continent of Australia is normally associated with the opal, but the Aussies are also proving to be a major force in the blue-sapphire market, and the fancy-sapphire market as well. They are certainly capable of producing lovely blue specimens, though some are in reality quite dark. These gems are sometimes called "midnight blue" or "ebony" sapphires. Similar colors from Thailand and Nigeria are also often seen at retail. Although many dismiss their color as being too dark, these midnight blue gems have taken some hold in the marketplace, and they offer the average consumer an opportunity to own affordable solitaires of considerable carat weight.

Besides the major sources highlighted already, others continue to produce fine specimens. Countries such as Cambodia, Colombia, Myanmar, and Brazil are all considered factors in the marketplace. In fact, some of those coming out of the ruby-rich regions in Myanmar are particularly beautiful, often mistaken for the cornflower blue variety from Sri Lanka. In addition, large-scale production of high-quality blue corundum began in Laos toward the end of 1996. In fact, many of those working the mines in Laos formerly mined rough in Kanchanaburi, and most of the mining equipment used in Laos has been transported from those sites in Thailand. In Africa, Rwanda was an increasingly important source, until the recent political turmoil there; gem experts also are excited about the quality of rough coming from Madagascar, where an incredible blue-sapphire rough of more than ninety thousand carats was recently found. These high-quality blue gems were discovered on the northern part of the island, and some contain inclusions of zircon.

On the home front, sapphires of many colors from Montana have staked their claim in the marketplace. Although they may be found all over the state, blue sapphires from Yogo Gulch are the only Montana blues that do not require heat treatment for color or clarity. They were once Montana's best-kept secret, but an aggressive marketing program has caused the rest of the world to take notice. Still, production of the Montana sapphires is small and sporadic compared to world standards, and it is doubtful that they will ever have a major impact on the sapphire market.

So which one do you buy? Once again, I must say that beauty is in the eye of the beholder. About the best advice I can pass along is that you should not buy a stone that is too light and loses its color, or one that is too dark and totally devoid of any blue color. The rest, my friend, is up to you!

Star Sapphire

As with other gems displaying the phenomenon called asterism, the star sapphire results when the needle-like inclusions called rutiles intersect at just the right point. The star sapphire, too, is almost always seen cut *en cabochon* and polished to a glossy finish. It is recognized as one of the birthstones for December. Although the most common color of star sapphires is blue, they are sometimes seen in others, including yellow, green, pink, gray, and lavender. No matter the color, natural star sapphires are rare and can be quite costly. For a long time, the star effect of the sapphire and the ruby was considered to be a trademark of nature that humans could not duplicate. Eventually, however, the research-and-development team of a major corporation developed a method of reproducing the phenomenon, and this synthetic stone is usually known today as the Linde star sapphire.

Among the impressive gems on display at the Smithsonian Institution in Washington, D.C., is a star sapphire of more than fifty carats hailing from the deposits of Sri Lanka. What makes this stone even more unusual is its vivid purple color. As you might suspect, the star sapphire is found just about anywhere there are large deposits of sapphire rough. The star sapphire is a gem that is not considered "overly feminine"—whatever that means—in appeal, and as such it has been a favorite for men. In fact, if you are looking for a gift for the man in your life, a star-sapphire ring would be an excellent choice.

Fancy Sapphires

By definition, a fancy sapphire is one that is any color other than blue. As with the more widely known blue variety of sapphire, the other colors of such corundums are sometimes indicative of their country of origin. Fancy sapphires are formed when iron combines with other trace elements such as chromium, titanium, and vanadium. If you thought up to now that all sapphires were blue, read on; it's time to learn something new! For ease of identification, we will describe each of the fancy sapphires by color or place of origin.

Tanzanian Fancy Sapphires

An exciting new development in the fancy-sapphire field is the recent discovery of multicolored sapphire rough in Tanzania. The initial material was found in the Songea region during the summer of 1994, and another alluvial deposit was found a bit later around the Muhuwesi River. Both sites are in the southern part of the East African country. Other minerals were found in the two locations as well, including chrysoberyl, quartz, spinel, zircon, and various garnets, among them tsavorite, the rare, emerald green grossular garnet. Fancy sapphires also have been found in alluvial deposits along the banks of the Ruvuma River, which separates southern Tanzania from Mozambique.

These colorful specimens have just recently reached the consumer marketplace in the United States. Found in a multitude of colors, the stones are 100 percent natural and, like other corundums, derive their color from iron and other trace elements. Many of the initial pieces of jewelry entering the retail market have taken the form of multicolored bands and clusters of striking beauty. Besides showing the entire colorful spectrum at once, the individual stones in these multicolored pieces are easier for the manufacturer to match.

All Tanzanian sapphires are found in pebble size and, as such, are generally not available in large solitaires in any color. Many of these stones are extracted by natives in primitive fashion, primarily from riverbeds, under natural waterfalls, and in areas of treacherous rapids; no deep-shaft mining is being done. If you see a multicolored band or cluster that catches your eye, definitely grab it! Remember, these creations will go with just about everything in your closet.

Color-Change Sapphire

Another exciting gem that has surfaced from this very same region of Tanzania is a variety of corundum that the trade simply calls a color-change sapphire. At first glance, this fascinating gem somewhat resembles rhodolite, a member of the red-garnet family. When exposed to direct sunlight, the stone displays vivid blue colors, along with a presence of red and in some instances even green and yellow. It is important to note that this color-change phenomenon is 100 percent natural, not a product of heat treatment. In fact, if the stone were to be heated, it would lose its color-change property. If you should happen to own one that needs to be sized or repaired in any way, warn your jeweler of its aversion to heat, since a carelessly used jewelers torch could ruin the stone forever. For this same reason, I would not recommend using steam or hot water to clean a color-change sapphire or other gemstone with such properties.

If there is a downside to this beautiful gem, I suppose it would be its typical size. Color-change sapphires are found in pebble deposits, much like the multicolored Tanzanian goods. As you can probably guess, this means the color-change variety is seldom seen in sizes weighing more than one carat. For those of you who demand a larger solitaire, synthetic color-change material has already been developed and is currently available at retail.

Green Sapphires

When the conventional blue meets the sparkling fancy yellow sapphire, they combine to produce much of what the world knows today as the green sapphire. The conventional green sapphire is described by most as bottle green, and this gem is often confused with the green tourmaline of a similar shade. To make things even more confusing, when the yellow color predominates, it results in a green gem that is a dead ringer for the peridot. The Chanthaburi mine in Thailand is a major source of this variety of corundum. When the blue predominates, the result is a lovely teal green color, found mainly in Sri Lanka. Overall, most green sapphire comes from Sri Lanka, Thailand, and Australia. Green sapphire is a birthstone for July.

Lavender Sapphire

When the chromium-rich pink sapphire is invaded by deposits of iron and titanium, the end result is a pinkish purple gem known as a lavender sapphire. The depth of the purple hue depends directly on the ratio of these elements, and this combination has produced some beautiful specimens. Colors vary from an almost regal purple through a light pastel violet hue. Lavender sapphires come primarily from Sri Lanka and the Chanthaburi mine in Thailand.

Pink Sapphires

Pink sapphire is colored by minute quantities of chromium. As the chromium level increases, the color becomes darker and more vivid. As with many other gems, those that are darker are generally more valuable than those of a light pastel pink color. These beautiful stones are considered by many gem experts to be the closest relative to the padparadscha. Some consider the pink sapphire to be the stone of the heart. It is also recognized as a birthstone for October. The primary source of pink sapphire is Sri Lanka, although Myanmar and East Africa also contribute to the world's supply. Contrary to popular belief, poor-quality red ruby does not hit a certain level and then become pink sapphire. Pink sapphires are much more valuable than poor-quality rubies of pink color.

Silver Sapphire

The least known of all the colored sapphires, the silver sapphire comes from only two places in the world: Sri Lanka and Australia. These silver-blue gems are found only in small, waterworn-pebble deposits. Gems of more than one carat are virtually nonexistent. Oddly, the area in Australia that produces this seldom-seen variety of corundum is known as Emerald, a remote mining area about five hundred miles from the city of Brisbane.

White and Colorless Sapphires

Most people lump the colorless sapphire and the white sapphire together, using the names interchangeably. Although they share many common bonds, there are also subtle differences that separate one from the other. Since this is an area of confusion to many casual gem collectors, let's discuss this in greater detail.

Both varieties belong to the same mineral group, corundum, which means they share the same approximate chemical and physical characteristics. Since all colorless and white sapphires are found in alluvial deposits as water-worn pebbles, large stones of several carats are virtually nonexistent. When found at retail, which isn't often, they are likely to be seen in clusters and bands, or as accent stones surrounding other colored corundums. In addition, both hail from the gem-rich nation of Sri Lanka, and they are commonly known in the industry as "colorless Ceylon" or "white Ceylon" sapphires. These particular varieties of sapphire are birthstones for April. Astrologers link the white sapphire to the planet Venus.

Beyond all that, however, colorless sapphire is corundum in its purest state, totally devoid of undertones. Highly prized and much sought after, colorless sapphire is virtually impossible to find. As impurities begin to invade, the gem takes on an almost white undertone, and it is this undertone that separates the two. Interestingly enough, some white sapphire actually was once a pale blue color, not deep enough to be classified as Ceylon blue. These stones are sometimes heated to remove the blue color, and they can also give off a slightly blue tinge when closely examined. These subtle differences give each variety its own appeal to the consumer. If an attractive solitaire, band, or cluster should present itself, you would be wise to add it to your growing gemstone collection. Who knows when you'll get another chance!

Yellow Sapphires

If you are thinking it's been a while since you've seen a yellow sapphire, it's not just your imagination. Demand for this beautiful, sunny gem has reached epic proportions in the Orient, prompting dealers to divert the stones to these markets. Needless to say, the increasing demand has resulted in higher prices for the yellow sapphire, and the U.S. market for the stone has for the moment pretty much dried up. One reason for the increased demand is that many people in the Far East associate it with emotional healing and increased energy levels. Astrologers have long since associated it with the planet Jupiter; it is also linked to Mercury. Sources of supply today include Sri Lanka, Thailand, Australia, and Cambodia. The state of Montana produces a vibrant yellow variety, in small sizes, that is now being seen in some local retail markets.

Feldspar

Although feldspars literally cover the Earth, as a group they remain almost unnoticed at the retail level by the average gem buyer. This section will delve into some of the group's most popular members. Since feldspar is so widely available, its future is obviously not in any doubt. Certain varieties are gaining in popularity among gem collectors, however, and that could someday pressure a particular variety of feldspar toward extinction.

As in any other mineral group, certain subspecies, including spectrolite and some varieties of moonstone, are rarer than others. As is the case with all other gems, an increase in the popularity of these different forms of feldspar will certainly result in higher retail prices down the road. For now at least, these fascinating gems are still within the average person's financial reach. Amateur gem buffs should selectively collect these various feldspars while they remain available and affordable.

Labradorite

Labradorite is the most popular member of the plagioclase family of feldspars. A most intriguing gem that can be faceted or cut *en cabochon* or flat, labradorite is found in many colors, which often serve to disclose the specific source. A gray variety that sometimes displays flashes of blue, purple, and even yellow originates in the peninsula of Labrador, the eastern Canadian site of its very first findings. The yellow variety usually is a native of the colder climes of the former Soviet Union. There is also a beautiful blue, semitransparent variety from Finland called spectrolite, worth its own separate discussion (see page 83). The ability of these types of feldspar to show varying intensities of color is a phenomenon known in the jewelry trade as labradorescence. Labradorite is also known as one of the "seven gems of winter"; I'm not exactly sure why, but I'll bet that if you spent most of your life buried in the frozen tundra, you'd probably find out. Suffice it to say that you won't see me riding a dogsled into the wintry climate of Canada or Finland to search for labradorite anytime soon.

Moonstone

A distinctive member of the orthoclase family (see below for other members), moonstone is perhaps the most famous feldspar of all. Unquestionably, it is one of the most intriguing gems in the world. Good-quality stones emit a white or bluish silver haze around the perimeter of the stone, giving it an adularescence like that of no other gem exhibiting this phenomenon (including its close relative the adularia, another orthoclase; see the next section). Ancients referred to the moonstone as "the pebbles of the moon," and it was said to warn of impending danger when held skyward during a full moon. In earlier times a full lunar eclipse struck terror into people's hearts, and they often blamed the moonstone for the disappearance of light. When the moon was seen at its fullest, ancients believed the gods were angry with them for disdaining the moonstone. Those born in April, June, August, or October may claim the moonstone as their birthstone. It is also accepted by those born under the sign of Cancer or Sagittarius as an astral gem.

A deposit of multicolored orthoclase called rainbow moonstone has entered the marketplace in the past few years. These gems have been said to come from a deposit in southern India. The find has existed since the late 1980s, but only recently have dealers considered the extraction method economically feasible. Since this is a relatively new variety of moonstone, little can be said about its long-term prognosis. To date, finds of stones of five carats or more have been isolated and sporadic at best.

High-quality moonstones of large proportion are rare and can be quite expensive. Much of the moonstone found at retail today hails from the gem-rich island of Sri Lanka, together with its neighbor, India. Other sources include Brazil, Myanmar, Madagascar, and Australia. In the United States, moonstone is sometimes found in Oregon and Virginia. Florida lists the moonstone as its official state stone (though it does not produce the gem). All you have to do is catch the full moon as it reflects off the sparkling waters and glistening sands of Florida's pristine white beaches to understand why.

Other Orthoclases

Sometimes referred to in history as "the gem of the noble," the orthoclase (a family that includes the moonstone) can be found in various shades of yellow and is often confused with labradorite. But that variety of feldspar (see above) has a higher specific gravity, meaning that an orthoclase will appear slightly larger when set side by side with a labradorite of identical cut and carat weight. The yellow orthoclase is found in Sri Lanka, Myanmar, Madagascar, and Brazil.

Another orthoclase that is coveted by collectors worldwide is a beautiful clear variety known as adularia, which exhibits a mystical-looking blue halo. This adularescence takes on a different sheen from that of its cousin the moon-

stone, since the adularia is a colorless gem. The rare adularia is found primarily in the majestic mountains of Switzerland.

Spectrolite

A little-known blue variety of plagioclase feldspar (the family that includes labradorite), spectrolite was discovered in Finland by a young lieutenant during World War II when an explosion from a land mine exposed a vibrant blue flash that would later be identified as the gem phenomenon known as labradorescence. As fate would have it, the lieutenant's father was head of the Geological Society of Finland at the time, and it was he who first named the stone spectrolite. Most gem historians now believe that spectrolite existed in the area for thousands of years before its discovery, having been forced deep into the Earth during the Ice Age.

True spectrolite that exhibits the phenomenon of labradorescence is found only in Finland, in the small mining region around Ylämaa, a little town also known as the Gem Village of Finland. This charming village is located just five miles from the Russian border. A source in Russia, near Saint Petersburg, and the island of Madagascar also contribute to the supply of spectrolite, although most gem authorities consider stones from those places to be more closely related to the moonstone, and the latter is sometimes known as Madagascar moonstone.

In order to display its inner glow, spectrolite is almost always seen cut in a slightly domed, flat manner. This inner glow can best be compared to a searchlight as it pierces a deep blue evening sky. It is a truly fascinating gem, seldom seen at conventional retail outlets. I would highly recommend that every casual gem buff add the spectrolite to his or her collection if the opportunity presents itself.

Sunstone

A relatively new entry just starting to make its mark at retail, sunstone belongs to the oligoclase family of feldspars. Usually seen cut *en cabochon* or flat, this unique gem can show shiny little iron-rich particles that look like glitter. These particles are generally rust or bronze but also can be red, orange, golden, silver, or green. The sparkling effect from these tiny particles is a phenomenon sometimes known as aventurescence. Traces of iron turn the sunstone to yellow, while reds are influenced by copper.

Although sunstone from the American Northwest has surfaced in local retail stores, it is still quite difficult to find in other parts of the nation and the rest of the world. This is unfortunate, since it has a look all its own. Sunstone is by nature dichroic or trichroic—that is, it may appear green and red, or green, red, and yellow, when seen from different angles. The material from Oregon may even exhibit the color-change phenomenon, showing red under natural light conditions and green when examined indoors.

Although the most highly prized gems are those with a vivid red hue, all sunstones are sights to behold. Stones can reach sixty carats or more, but most of the rough has generally been running between fifteen and thirty carats. The largest sunstone found to date is a spectacular gem of nearly three hundred carats; unfortunately, stones of this size are turning out to be the exception, not the rule. Sunstone is also sometimes known as aventurine feldspar, which causes it to be confused at times with a chalcedony known as aventurine quartz. In addition, consumers should be aware of a man-made glass deliberately included with copper and called goldstone. This artificial material is sometimes passed off to unwary shoppers as sunstone.

For many years, the material coming out of Oregon was available on a hit-or-miss basis, and as a result, consumer demand was low. However, new cost-effective mining methods are finally starting to push this fascinating gem into the limelight. In fact, sunstone already surpasses both tourmaline and peridot in volume of exports. Sunstone can be found in Norway, Denmark, Canada, and parts of the former Soviet Union, but so far the goods coming out of Oregon control this rather limited marketplace.

Between its glittering copper, its di- and trichroism, and its color-change phenomenon, sunstone is truly one of the most appealing gems of all. For budget-minded gem buffs, the costliest sunstones are those in the red, pink, and orange family, generally speaking, while those in green, yellow, and green-yellow combinations tend to be a little less expensive. Additionally, sunstone may appear in a colorless state, but this variety is actually the least sought after of all. If supplies remain consistent, aggressive marketing efforts could bring sunstone into the mainstream retail jewelry market in the not-too-distant future. If you see one that fits into your budget and catches your eye, by all means go for it.

arnet

One of the most interesting mineral groups of all is the garnet. To the uninformed general public, the garnet is a brownish red stone of no particular importance, but gem enthusiasts know there is a lot more to this colorful family than that. For example, did you know that most garnets are not artificially enhanced in any way? This alone should put the garnet in a class all by itself. For the most part, the garnet is a much-maligned, underrated gem.

For now, the world's largest supply of garnets is in the continent of Africa, but a major new find might change that someday. In December 1996 in central Australia, a university student unearthed what is believed to be the largest single garnet find ever: this mammoth rough is thought to weigh thousands of tons and measure nearly one hundred feet across!

Like any other mineral group in abundance, the garnet group will remain in (and deep within) the world long after you and I are gone. Still, certain family members can be considered rare even today. I have always felt that the garnet group as a whole is somewhat undervalued, and that everyone who can should collect all the various species in one size or shape while they are still affordable. If this is not financially feasible, keep an eye out for jewelry that contains more than one variety of garnet, such as almandine with pyrope, or tsavorite with mandarin (a variety of spessartine).

Since this is such a large and varied group, it can probably be said that some kind or another of the garnet mineral group can be found pretty much anywhere in the world. Some of its most significant sources include India, parts of the former Soviet Union, and Brazil, in addition to the previously mentioned finds in Africa and Australia. Domestically, garnets can be found in many states, including Arizona, California, Idaho, Maine, Montana, New Mexico, and North Carolina.

The Red-Garnet Group

Spend a few days at a gem-sorting table, and it quickly becomes apparent that there are a hundred shades of red. Never was this truer than in the case of the red-garnet group. The shades of red that define the various types of garnet (to the naked eye, at least) are caused by the presence of iron, chromium, and manganese that invade the crystal. As long as each keeps to itself, everything is just hunky-dory. The problems start when they decide to grow together and borrow little bits of this and that from each other, making identification and separation of each a daunting task to the everyday consumer.

For example, a garnet classified as almandine may borrow a minute amount of chromium from the pyrope, and thus be a little more red and a little less brown than an almandine garnet from another deposit. One of my favorite red garnets, from the East African nation of Mozambique, is a classic case in point of an almandine garnet that at times almost resembles a ruby to the naked eye. Each individual circumstance can be different; each and every stone may vary slightly. If this all sounds a bit confusing to you, take heart, for you're not alone. I have sold hundreds of millions of dollars' worth of colored gems over the years, and to this day I, too, have a lot of difficulty separating one red garnet from another.

Thankfully, a trained gemologist or jeweler can reduce this formidable task to a simple process of identification. These different red garnets vary in hardness and specific gravity, which makes it easy for an expert to separate them with just a few tests.

Almandine

The almandine is the most commonly seen member of the red-garnet family. Its color is usually deep red, sometimes almost appearing to be black when viewed from a distance. Found in large deposits worldwide, the almandine owes its deep color mostly to the presence of iron. Just how red the almandine is

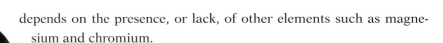

depends on the presence, or lack, of other elements such as magnesium and chromium.

The garnet is the most widely accepted traditional birthstone for January (though other traditions consider it a birthstone for February, May, or October), and for more people born in that month, the almandine is their garnet of choice. Astrologers generally see it as the traditional astral stone for Aquarius, though some attribute it to Leo or Capricorn. Astrologers also regard the garnet as a strong, positive gem, bringing true happiness to those who wear it. Ancient people believed the garnet could bring relief from debilitating arthritis; some thought that when a wise man had a dream about garnets, he would soon reach a solution to the mysteries of life.

Pyrope

Pyrope is the bright cherry red garnet, colored by chromium. These garnets are usually found in areas of volcanic rock or within waterworn pebbles in alluvial deposits. A small pyrope garnet found in Arizona is sometimes called the anthill garnet. As you might guess, these bright red specimens have been observed in anthills that are approximately the size of Mount Everest, as a result of constant excavation by these tiny miners. (To better understand this, you have to take a closer look at ants, which in Arizona and throughout the American Southwest are about the same size as small cattle. Spend an hour just watching ants and you will arrive at the same conclusion I have: the ant is a pretty stupid insect. What else can you say about anything that spends its entire life working, working, working—only to end up squashed like a grape for its trouble?)

The pyrope, like the almandine, has its share of folklore and superstition. Ancients believed that its deep, bloodlike color would protect the wearer from injuries and blood loss; warriors often carried it into battle. Since earlier astrologers did not have the tools of detection available today, many gems that were thought then to be rubies were actually pyrope (and, to a lesser extent, almandine) garnets. Because of this, the pyrope was often considered among the most powerful gems in the universe, as well as a gift from the sun. Some regard it as an astral stone for Capricorn, and astrologers link it to the planet Mars.

Although it can sometimes be confused with the almandine, the pyrope should be red, with no undertones of brown whatsoever. Like its cousin the almandine, the pyrope is also found pretty much worldwide. Incidentally, because pyrope is sometimes found in areas of carbon deposits, its presence can indicate a diamond find in the vicinity. News of an exciting new find in the American Southwest has recently been reported, and these gems are said to have some of the highest concentrations of chromium anywhere. If true, this will classify them as some of the deepest red pyropes ever found. Unfortunately, these gems have turned up in very small sizes, making large solitaires almost nonexistent. Other nations with large concentrations include several in Africa, as well as Australia,

Brazil, and Myanmar. Unlike the almandine, pyropes of heavy carat weight are seldom found at retail outlets, and they tend to be quite costly when they are.

Rhodolite

Rhodolite is a variety of red garnet found most often wherever the almandine and the pyrope occur. It is neither almandine nor pyrope in nature, but rather a little of each. Although each stone varies slightly, its mixture is normally about 55 percent pyrope and 37 percent almandine, along with other elements. Its color has been described as raspberry, but it can be more pink or purple depending on the mixture; raspberry is generally the color of choice for garnet collectors. This lovely gemstone is found mostly in the gem gravel of Sri Lanka, in small deposits of waterworn pebbles. Because of this, large stones of five or more carats are considered rare. In addition to Sri Lanka, most rhodolite today comes from Thailand, Brazil, India, and several countries in Africa. In the United States, North Carolina is generally considered its most important source. Rhodolites from Thailand are similar in color to the violet rubies that can sometimes be found there today, although these two gems have nothing in common other than their country of origin. Since the rhodolite and the ruby are not of similar structure, a trained gemologist or jeweler can easily tell one from the other by putting the gem through some simple tests.

I personally believe the rhodolite should be a part of every gem collection, for it is most beautiful and unusual. Although large solitaires can be quite costly, more affordable smaller gems of one or two carats are generally within the consumers' reach. For those even more budget-minded, rhodolites (like other red garnets) are sometimes encountered in sterling silver or gold plate.

Star Garnets

When visible inclusions intersect at precise angles in an almandine, these interior irregularities result in a little-known gem called the star garnet. The inclusions are most often black, and the stone itself is usually a purple-red to plum color giving off an almost blue sheen. As with other gems exhibiting this phenomenon, the stone is always cut *en cabochon* and highly polished to accentuate its asterism. Although star garnets are not generally high-end gems, they are considered quite rare and should be a part of your gem collection. Star garnets can be found virtually anywhere the almandine occurs worldwide. In the United States, Arizona, Colorado, Idaho, Maine, and North Carolina are all locations with the potential of finding the star garnet.

Red-Garnet Wrap-Up

Remember that all of the red garnets we have examined to this point may be found in fascinating combinations with others. Almandine-pyrope, almandine-spessartine, pyrope-spessartine, and other examples bring new and unusual shades of red to the garnet gem group, and I suppose technically they should all be considered red garnets. The lesson to be learned from this is that all red garnets are not created equally. As with other gems, the red portion of the spectrum is as wide as it is long. The garnet lover may own two, three, or more of these gems, none of which share the same shade of red. If this really bothers you, feel free to have them tested. To me, however, every gem has its own "personality," and this is what makes collecting red garnets that span the whole range so much fun to begin with.

Malaia Garnet

Red garnets from Tanzania and Kenya that contain rutiles (needle-like inclusions) are commonly known as Malaia garnets. These gems are actually a mixture of almandine, pyrope, and spessartine. The Malaia garnet does not fit into any of the other red-garnet categories; in fact, its name literally means "not of the family." The rutiles that produce the fascinating effects in these vivid red garnets are the same kind that produce the golden needles seen in rutilated quartz. They are also responsible for the cat's-eye and asterism found in many gems worldwide, including corundum, quartz, and chrysoberyl, as well as garnet.

Spessartine Garnet

Spessartine is an orange to red-orange member of the garnet mineral group that is primarily colored by iron. Until recently it has been nearly impossible to spot at retail, but important new finds promoted by TV shopping networks are now slowly bringing this beautiful gem to its adoring public. Astrologers link spessartine to Rahu, the northern lunar node, where the sun and moon appear to cross paths in the celestial sphere (resulting in eclipses and, according to some cultures and astrologers, unexplained phenomena).

Mandarin Garnet

The mandarin garnet, a new variety of spessartine, is just starting to make its presence known in the retail market. Mandarin garnets were first discovered in

the gem-rich African nation of Namibia, followed by a second source in Pakistan. There, in the unforgiving Himalayas in the Pakistani region of Kashmir near the border of India, these garnets have a short and dangerous mining season. These brightly colored orange gems are sometimes found in conjunction with green tourmaline and quartz crystal.

Mandarin garnets first made their appearance on the scene during the 1993 gem show in Tucson, Arizona. Although the mandarin was initially slow to receive dealer acceptance, demand is expected to increase dramatically as news of its discovery makes its way across the United States. Since ones of a deep red color are nearly impossible to get, the mandarin garnet is most often seen in a range from red-orange to orange. If you do find a mandarin garnet, grab it, because its future supply is questionable at best.

The Colorful Grossular Garnets

From the first discovery in Russia of a distinctive green garnet to the most recent apparent find of a few blue garnets in East Africa, the grossular family has found its niche in the world of gems. Members of this family include the tsavorite, a vivid green gem that can be confused with the natural emerald, and the multi-colored hessonite, usually observed in the jewelry world as a brown garnet with some orange undertones. The grossularite, a gem that shows more yellow than green, and the rosolite, a little-known pink specimen found only in Africa and Mexico, are other grossular garnets that are more difficult to find. Besides these colorful varieties, grossular garnets can be found at times in a pure, colorless state. In addition, the unconfirmed report of a rare and exciting new discovery in eastern Africa produced a few blue garnets attributed to the grossular-garnet group. These were said to be the first of their kind on record. Unfortunately, the deposit was reportedly quite small and isolated. Needless to say, these history-making gems never made it to retail.

Everyday gem buffs sometimes confuse various grossular garnets with natural and synthetic spinel. If you are ever in doubt as to the composition of your own gem, remember that the natural one has a different refractive index, and a quick test for fluorescence and inclusions will separate it from the synthetic material.

Grossularite

Some people may believe that all grossular garnets are green, but that is simply not the case. When yellow and green combine, the gem is known as a grossularite garnet. Inexperienced consumers have often mistaken the grossularite for the peridot. Sri Lanka, Pakistan, parts of the former Soviet Union, and countries in Africa are all important sources of the grossularite. Africa also produces a semi-transparent grossular garnet known as Transvaal jade or African jade, which does in fact resemble gem-quality jadeite. When found in jewelry, Transvaal jade is usually cut flat or *en cabochon*. The gem also occurs in massive forms and is sometimes even fashioned into works of art.

Hessonite

The hessonite garnet is found by sifting through gem gravel, rather than by using deep-shaft mining as with some other grossular garnets. Translated from the Greek, the name means "inferior," which has influenced the novice gem consumer to avoid this unusual golden brown gem. This is unfortunate, because the word *inferior* does not reflect the stone's general quality; instead it refers to the gem's hardness, which is rated 7.25, compared to that of its look-alike rival, the hyacinth zircon, which has a hardness rating of 7.50. Hessonite gets its cinnamon-like color because of the presence of iron and manganese in the crystal, and in the gem world hessonite is sometimes known as cinnamon stone. Astrologers have linked it to Rahu, the northern lunar node, where the sun and the moon appear to cross paths in the sky. Like the tsavorite, it is found primarily in alluvial and other secondary deposits, making gems of large carat weight quite difficult to find anywhere. The best sources of hessonite include Sri Lanka, Brazil, and Russia. In the United States, it is most likely to be found in Maine.

To those consumers who wear a lot of earth tones, hessonite would make an excellent addition to your wardrobe. To others, its brownish color might be considered less attractive, particularly when compared to the bright green tsavorite. Still, hessonite is seldom seen at retail, and it would be a worthwhile purchase for anyone in the process of gathering a gem collection.

Tsavorite

Perhaps no grossular garnet is as well known as the tsavorite (pronounced "sav-oh-rite"), a beautiful emerald green gem found primarily in Kenya. Its quality may be so great that it may be classified as flawless. Tsavorite is sometimes found in rocks that remind me of potatoes—baked, not fried. It was initially discovered along the banks of the Tsavo River in southeastern Kenya. Tsavorite obviously gets its name from the river and the surrounding area, a wilderness preserve of over eight thousand square miles known as Tsavo National Park. Tsavo is one of the largest national reserves in the world and offers breathtaking views of Mount Kilimanjaro in neighboring Tanzania. First located in alluvial deposits along the banks of the river, tsavorite is found these days mostly in small pockets scattered throughout the national park.

Although small deposits of tsavorite have occasionally surfaced in other parts of Africa, including recent finds in Tanzania, Kenya is responsible for nearly the entire world's supply of this gem. Deposits are at best irregular and sporadic, and the future of tsavorite is always in doubt. Since 1968, prospectors have discovered more than fifty deposits, yet only one is still considered a reliable source of tsavorite. Some astrologers link tsavorite to the planet Mercury.

Colored by trace elements of vanadium, the tsavorite garnet can at times rival the highest-quality emeralds in the world. Although it is an alternative to the emerald, it is hardly considered an inexpensive look-alike. This may come as a surprise to many consumers, who look on the garnet family as a whole as being affordable. This is true of certain varieties, but the tsavorite isn't one of them. Because it is most often found as pebbles in alluvial deposits, stones of three carats or more are just about impossible to find. Such stones quickly become the property of wealthy gem investors. I once observed a two-carat solitaire at a gem show with a price tag of over $3,000—wholesale! Obviously, you should grab anything that even resembles a solitaire—assuming, that is, that you can find one anywhere.

Andradite Garnets

The andradite garnet family is a lesser-known yet colorful group of gems. This group includes the demantoid, a bright, vivid emerald green stone found in parts of the former Soviet Union; the topazolite, a yellow gem found mostly in pebble-size deposits; and the mysterious black melanite, which sometimes also is found in very deep shades of red.

Demantoid

The most sought-after member of the andradite garnet family is the demantoid, which often can be found in serpentine rock. This stone generally contains curved fibers of asbestos, inclusions known in the gem world as horsetails. It is sometimes confused with low zircon, yet it can be costlier than tsavorite and even harder to find. The first demantoid garnets ever discovered were unearthed in the Ural Mountains of Russia in the mid-1800s. The gem was first thought to be chrysolite (known today as peridot) and was not classified as a garnet until ten years after its discovery. Eventually named andradite, the gem was a hit with the Russian retail jewelry trade for quite a while. Beautiful pins and other objects of art fashioned out of demantoid were very popular during Victorian times. Today the demantoid garnet is mostly seen in pieces of estate jewelry and antique collections. This gem can fetch upwards of $3,000 per carat when found in sizes weighing three carats or more.

New deposits of demantoid have turned up in Russia, and initial reports claim these will yield some of the highest-quality demantoids ever found. New finds have also surfaced in Namibia, although their quality has not yet been determined. In general, Namibia produces some of the highest-quality diamonds and colored gems in the world, and experts believe there is no reason to suspect these demantoids are anything less.

Besides Russia, the most significant current sources of demantoid are Kenya and Congo. During my exhaustive research for this book, I found that the demantoid garnet for some reason is associated with the name Dennis, which has an obvious connection for me. That I choose to include this information testifies to the fact that there just isn't a whole lot of information in the world about the andradite group.

Uvarovite Garnet

The winner of the hard-to-pronounce contest is the uvarovite (pronounced "you've-ahr-uh-vite"), a colorful garnet that was first discovered in the Ural Mountains of Russia. In fact, at one time it was known as the "emerald of the Urals." (If you already knew *that* obscure piece of trivia, if I were you I would call the folks at *Jeopardy* as soon as possible.) Uvarovite can also be found in Turkey (which makes it an excellent gift for Thanksgiving). Italy and Finland are two other sources of this elusive gem, and California and Texas produce sporadic supplies that nobody cares about anyway.

Uvarovite is generally bright green, a good indicator of its chromium-rich nature. Because it is largely regarded as a brittle stone, extremely difficult to cut and facet, this variety of garnet is seldom seen at retail. In twelve years I have seen literally hundreds of types of gemstones come and go, but I can recall seeing just one uvarovite in all that time (and not even one other gem that begins with the letter *U*).

Lapis Lazuli

Although lapis lazuli (pronounced as if it's one word, with the emphasis on "slaz"), also called just lapis, can be found in Chile, parts of the former Soviet Union, and the United States, the only material considered to be of gem quality comes from the mountains of Afghanistan, where it has been mined for thousands of years. Consumers generally think of lapis as a bright blue gem, but it also can be found in shades of red, purple, and black, though none of these colors are classified as gem-grade material and so those kinds are found only at gem and mineral shows, pretty much the private property of the rock hound.

Lapis lazuli is a complicated gem with a structure that can vary considerably. Its chief component is a mineral known as lazurite, which accounts for up to 60 percent of its structure. Lapis sometimes seems to sparkle, which indicates the presence of pyrite, also known to consumers as fool's gold. Other components in lapis include calcite and sodalite. The ratio of these materials to one another can affect the hardness, specific gravity, and even the color of lapis lazuli.

Like many other ancient gems, lapis lazuli has a long and storied past and a place of importance among gem collectors. Those born in September or December may make claim to lapis as their birthstone. A stone of much cosmic significance, lapis is often considered an astral stone for Aquarius and Taurus. Ancient astrologers linked lapis to the planet Saturn; some link it to Venus. It was a practice of certain ancient Egyptian cultures to bury the dead with a lapis scarab for protection. In fact, the earliest cultures actually valued lapis more than they did gold. Greeks once spoke of an ancient sapphire included with gold, which was undoubtedly lapis. Some believed that dreaming of lapis would foretell the arrival of a forever-faithful love. Medicinally, it was once thought to be an aid against fevers, sore throats, and burns. In Victorian times, lapis was sometimes ground into powdered form and used as a pigment for elaborate artifacts and brooches of glass and enamel.

Although lapis is essentially a sole-source gem, there is little concern as of now about its future supply. This is as much a result of lack of demand as anything else. Even with its historical significance, most consumers are not aware of its lofty status in the gem world. To this day, lapis is next to impossible to find at conventional retail outlets. However, exposure on nationally televised shopping networks has begun to better educate the general buying public, bringing about increased demand for lapis. Despite the relatively low cost of natural lapis, simulants also exist. In fact, certain chalcedony agates can be dyed blue and passed off to unsuspecting consumers as lapis lazuli. Because of its long history, I believe even the most casual gem buff should include at least one piece of lapis lazuli in his or her collection.

Opal

Generally speaking, opal is divided into two categories: common and precious. Common opal is of little value and is not considered to be of gem quality, so it bears nothing more than a mention here; precious opal, however, is of great importance in the gem world. For the purposes of this discussion, we will divide the precious opal into three groups: white (which dominates the opal market overwhelmingly and so will be discussed first), black, and fire opal.

White Precious Opal

If you own a white precious opal, you possess a gemstone that exhibits a gem phenomenon: in this case, color play, a colorful effect caused by diffraction. As light enters the opal, it bends around the edges of tiny particles of hydrated silica, spheres (or "chips") of silicon and oxygen suspended in water within the stone. When it is diffracted, the light (which is made up of all the visible colors, each with its own wavelength) produces an entire rainbow of colors.

The white precious opal is composed of silica spheres rather than a crystal structure, like most minerals. Technically speaking, opal is classified as a noncrystalline (or amorphous) variety of quartz, which makes it a distant relative of gems such as the amethyst and the various chalcedonies. White precious opal is sometimes seen in combination with other colored gems that produce a color-play effect, while diamonds (or sometimes cubic zirconia) are often used to frame the magical colors at play.

Australia produces approximately 95 percent of the world's supply of white precious opals. Much of the continent is rich in opal-bearing land, but only certain portions have enough deposits to justify the great expense of opal mining. Lightning Ridge, a mining town located about ten hours by car northwest of Sydney, is generally regarded as the largest opal-bearing site in the world. This area, known for frequent strikes of lightning that hammer its ridge, has been the home of some of the most famous of all opals, including Big Ben, a find of over

800 grams (more than 4,000 carats), and the spectacular 450-gram (2,250-carat) Light of the World, which some in the gem world consider the most colorful white precious opal ever found. Another famous opal-bearing region is Coober Pedy, a desert land so harsh, hot, and dry that its inhabitants live primarily in underground dwellings. Opals in Australia are plentiful, and devoted rock hounds delight in their abundance. If you've never prospected for opals before, take a guide or other experienced person with you when you go. Opal mining is serious stuff, not for the first-time visitor; you need to know what you're doing or you could end up in trouble.

The white precious opal is the subject of much folklore and tradition. Because it contains every color of the rainbow, early peoples felt it was a gift from the heavens. They also believed the opal had magical powers and could heal the sick and bring good fortune to those in need. The white precious opal has been widely accepted as the traditional birthstone for October and is considered an astral gem for the signs of Libra, Scorpio, and Capricorn. The ancient Romans regarded the opal as a sign of loyalty and hope, while other cultures believed its fiery colors were caused by lightning striking the stones as they fell from the heavens. To dream of an opal was thought to predict opportunities to come.

Since you may already have some preconceived notions about the opal that may affect your buying decision, let's set the record straight right here and now. Some people believe the opal will cloud up when exposed to water. Although there have been reports of hazing as a result of the glue used in the production of a doublet or triplet, water does not affect genuine precious opal. In fact, some experts recommend an occasional soaking to keep the stone moist and prevent any cracks. The opal does not shrink over time, nor does it lose its color in bright sunlight or intense cold. And finally, the opal is *not* unlucky. That tall tale was started by diamond merchants during the early part of the twentieth century, in order to protect their territory by convincing people not to buy opals.

Most consumers have just one gem in mind when they think of the opal: the translucent white stone with sporadic bits of color play, seen commonly at retail. Certainly this is the most popular variety of white opal by far, but it is by no means the only type in existence. White opals that display these beautiful patterns of light in an almost checkerboard pattern are known as harlequin opals. The jelly opal, another variety of white opal occasionally seen at retail, is easily spotted because of its translucent state. Totally devoid of the color play found in other varieties, the jelly opal has a yellow-and-blue sheen that always somehow reminds me of the moonstone. In addition, a new variety of white-and-black-spotted opal has recently surfaced in Mexico. This unique gem is just beginning to take hold. Initially it was anticipated in the fourth quarter of 1996, but limited supplies delayed its debut until the second half of 1997. Since current supplies are unpredictable and the future of this new gem is yet to be determined, add one to your collection as soon as possible if it strikes your fancy.

Consumers will not have to fret anytime soon about the supply of precious opal from Australia. Much to the chagrin of the miners and mine lords, supplies are so good that prices have remained quite low over time. Unless a vast new market for white precious opal develops, that situation will most likely continue for the foreseeable future.

The Legend of the Oregon Opal

Another variety of white precious opal, found in the northeastern section of Oregon, is a most unusual gem known simply as the Oregon opal. Like the jelly opal, this little-known kind is translucent. The Oregon opal, however, is usually faceted, which makes it look (to me, at least) like white quartz. Legend has it that this gem was discovered near the end of the nineteenth century by a shepherd, who came across the gem quite by accident. According to the story, he was tending his flocks in an area known today as Opal Butte when his curiosity was struck by a most unusual rock. When he picked it up, he found it to be uncommonly light. Deciding to investigate further, he struck the rock on a boulder until it split in two. Inside was a white, glistening material that a local gemologist described as precious opal. In time, it came to bear the name of the state in which it was found. When word of this discovery got out, rock hounds came from everywhere to work the fields for Oregon opal. Eventually, the land rights were leased to a logging company and the area was closed to prospecting.

Today, for the first time, a small mechanized mining operation exists on the butte, and now the Oregon opal can occasionally be found at retail. Demand for this variety is small, since most consumers are not even aware of its existence. I suppose that's fine with the operators of the mine, since they probably would have trouble meeting a heavy demand anyhow. Besides the Oregon opal, the potato-like rocks (called geodes) there contain a wide variety of other minerals, such as quartz, olivine, chalcedony, and common opal.

Black Precious Opal

Black opal is actually a generic term given to any opal with a dark body color when viewed face front. True black opals of excellent quality and color are some of the most highly prized gems on Earth. Opals with a dark background that is not intense enough to be termed *black* are commonly known as semiblack. Dealers stretch the color to the limit, attempting to classify any dark opal as black rather than semiblack. The reason is obvious: the black opal will command a much higher price at retail than the semiblack variety.

The black precious opal is pretty much the exclusive property of Australia. As with the white precious opal, Lightning Ridge produces nearly the entire world's supply of this very rare gem. Its structure is similar to that of the white precious opal, but with a deep gray or black background. This dark backdrop is the perfect setting for the fiery colors this stone displays, making it one of the most cherished of all the natural resources in Australia. Small deposits of black precious opal are occasionally found in other sections of Australia, but until recently none were considered to be significant compared to the supply from Lightning Ridge. However, a new find of high-quality black opals has turned up in an area known as the Wyoming fields. Nearly 150 new claims have already been staked, and by now some have been classified as Lightning Ridge quality. Good-sized stones show green against blue, orange, or red. This material is said to have fewer inclusions than other opal traditionally found in this part of Australia.

A second variety similar in appearance to the coveted black precious opal is a gem commonly known as the boulder opal. This variety is mined in the state of Queensland, Australia, where ironstone boulders occur with thin deposits of opal that span their surface. These opals are used to create an effect called a natural doublet. Because the effect is a result of the opal face and the black ironstone backing, the boulder opal is generally accepted as a genuine precious opal. Although they are of considerably more value than conventional man-made doublets, boulder opals don't begin to approach the value of the black precious opal. Boulder opals should be clearly identified, for the average consumer has been taken in by less-reputable dealers before. If you encounter a boulder opal of excellent color at retail, by all means buy it if it fits your budget. It will prove to be a much better choice than the man-made doublet or triplet, and you won't have to win the lottery to afford one.

Since supplies of the black precious opal are pretty much confined to one location, this stunning gem will continue to enjoy its lofty status. Black opal lives on the brink of extinction, and if by any chance you should come across a stone that is within your financial reach, grab it before it becomes too late. Because of impostors, however, you should request vital documents that prove the stone in question is a genuine natural black opal, not a boulder opal, doublet, or triplet. Suffice it to say that if the opal in question costs any less than what you paid for your first car, it's most likely a boulder or man-made gem.

Fire Opal

If the casual observer were asked to identify the fire opal, he or she would probably have to make a calculated guess at best, because it looks nothing like its cousins. The fire opal is usually faceted, but it can also be seen *en cabochon*. This gem spans a range of colors from yellow to red, with the best stones usually showing a vivid burnt red-orange combination. Good-quality fire opals are clean and clear, while lesser grades tend to take on a cloudy appearance.

Fire opal is generally found in volcanic rock formations in an area of southern Mexico near the beautiful Yucatán Peninsula. This area so dominates the world's supply of fire opal that the stone is sometimes known as Mexican fire opal. Since a small quantity of fire opal continues to be found in a remote region of the former Soviet Union, this label should not be used unless the source of the gem has clearly been defined as Mexican. Thanks to careful control of the flow of fire opal into the marketplace, prices have steadily risen, as consumer awareness of this gem has increased.

Like the black precious opal, the fire opal pretty much depends on one area of the world for its supply; however, that's where the similarity ends. Even though the popularity of fire opal today is at an all-time high, the stone has yet to really find its niche in the marketplace. Consumer awareness, though improving, remains quite low and supplies are not threatened. As with any gem that is hard to find, once it finally does take hold the availability of fire opal could change drastically in a relatively short period of time. Retailers have already absorbed a number of price increases, and this trend will continue if demand does surge. If you come across a fire opal of deep orange or red, be sure to make it a part of your collection; it's not likely to get any less expensive down the road.

Peridot (Chrysolite)

Peridot is the name given to the gem variety of the olivine mineral group. Historically, one of the oldest and most important sources of peridot was the tiny island known today as Zabargad. Now the property of Egypt, this strip of land less than two miles square has been documented as the first source of peridot, dating back as far as four thousand years. The earliest Crusaders, who called the place Saint John's Island, introduced this gem (then known as chrysolite) to Europe on their return home from battle. This source was reestablished shortly after the start of the twentieth century, after being closed for nearly a hundred years. Declared off-limits to outsiders in 1958 after being nationalized as part of Egypt, this island controls just a small portion of the worldwide peridot market today. The peridot is a stone that is documented in many ancient references, most commonly as chrysolite. It is mentioned throughout the Bible, and early Christians believed the stone to be sacred. Even today the peridot, along with the amethyst, is part of the tradition of Catholic bishops, found in a ring that symbolizes purity and morality.

Much superstition and tradition surround the lively green peridot. People born in August, September, or November often adopt this lovely gem as their birthstone. A gem of great mystical connection, the peridot is linked to a number of astral signs, including Pisces, Gemini, Leo, Virgo, and Libra. Astrologers also link it to Mercury and Venus. The peridot is considered a stone of springtime, and ancients believed it was a gift from Mother Nature in celebration of the annual creation of a new world. Napoleon gave peridot to the empress Josephine as a symbol of his undying love and admiration—obviously at some point before he had their marriage annulled. National leaders who publicly wore peridot generally were held in the highest regard and thought to be gentle, fair, and wise. Certain Asian cultures once believed that placing a peridot under a person's tongue would keep him or her from dehydrating. (It's not true, however, that the very earliest Gatorade ever made contained ground-up peridot.) On the other hand, in ancient times a dream about peridot was believed to foretell impending danger, causing the dreamer to exercise complete caution the next day; sometimes this person was so stricken with fear that he or she would remain in bed, virtually motionless, for days on end until it was felt the danger had passed! Many of these unhappy dreamers would refuse food or even water, for fear that a poison would ravage their bodies.

Today the San Carlos Apache Reservation in Gila County, Arizona, is by far the most dominant source of gem-grade peridot. In fact, more than 80 percent of the world's supply comes from this dry desert area. Despite its significant presence there, large stones of several carats or more are seldom seen. Arizona peridot is usually mined in the cooler months, when the threat of desert wildfires is at its lowest point. Domestically, peridot is also sometimes found on the beautiful islands of Hawaii.

Other significant sources of supply include Myanmar, Brazil, Australia, Norway, and countries in Africa. The Chinese contribute a variety of peridot that many regard as the most brilliant of all. It is found primarily as small, waterworn pebbles, and gems of even a carat or more are considered rare and collectible. More recently, a peridot deposit has been found in Ethiopia. Although these gems are said to be of similar color and quality as those from Arizona, so far it, too, has been confined to stones of smaller carat weight.

Overall, peridot is not a rare gem, yet high-quality large solitaires of five carats or more are considered unusual finds, and they can be quite costly. This may change dramatically over time, for a recent find of yellow-to-green top-quality peridot was discovered in the early 1990s in the rugged Himalayas of India. This find is significant because of the size of the individual crystals. In fact, one such discovery has already yielded a cut gem of over three hundred carats! Although it is a relatively new source of the gem, dealers are hopeful that the Himalayan discovery will eventually make available stones of excellent size and carat weight, so much in demand at retail. This remote location experiences some of the harshest winters on Earth, however, and will prove a challenge to even the most seasoned gem explorer.

Before it became Saint John's Island, the original source of peridot was known to the ancients as Topazios. Because of this, the gem was often mislabeled as topaz rather than differentiated as olivine. As if this wasn't confusing enough, a form of chrysoberyl nearly identical in color eventually hit the market, and like peridot it was known as chrysolite. Eventually, the differences between topaz and olivine became clear, and the yellow-to-green variety of chrysoberyl was finally reclassified into its appropriate mineral group. The name *chrysolite*

was eventually retired, leaving peridot as the only name still properly used to denote gem-quality olivine.

Although the peridot is the only gem-quality olivine, it may come as a surprise to some that another variety is occasionally seen in nature. This rather unattractive brown stone gets its color from years of oxidation and weathering. This material, of no particular importance, is sometimes confused with certain low-end garnets, such as those used in the manufacture of garnet paper for industrial purposes.

Quartz

Quartz is a common mineral, found in many parts of the world. Natural quartz takes shape in a number of different ways, and it can be crystal clear, translucent, or opaque. It spans just about every color in the spectrum, some of which are encountered regularly when you are shopping for gemstones. Quartz cannot as a whole be classified as a family of rare gems, yet some of its colors are seldom seen at retail. Let's take a more in-depth look at this colorful and interesting group.

Amethyst

The lavender-through-purple variety of quartz is called amethyst. It is without question the most popular member of the quartz mineral group, annually ranking among the elite of colored gems in total sales volume. Amethyst is the traditional birthstone for February and the traditional astral stone for those born under the sign of Aquarius, Pisces, Aries, or Capricorn. Some gem astrologers associate the amethyst with the planet Saturn, no doubt because of its broad range of colors.

One property that makes amethyst such an interesting stone is its vast color range, and these various hues are often indicative of its source. It is pretty much common knowledge that the darker varieties of amethyst are considered the most valuable. Although many sources of amethyst found worldwide are capable of producing dark, high-quality gems, it is widely agreed that those from the gem-rich, fertile lands of Namibia and Zambia are the world's finest. Commonly known as African amethyst, this variety is sometimes so deep in color that from a distance it almost resembles morion, a seldom-seen black variety of quartz. In its best deep purple color, African amethyst often commands a considerably higher price than any other variety at retail, which causes consumers who are more budget-minded to consider other sources. Brazil is generally regarded as the most significant source of amethyst in the world today, and that is the variety you will most likely encounter at retail. The fabled Anahi Mine in neighboring Bolivia also produces many beautiful deep specimens, though not in nearly the same volume as Brazil. Amethyst from Russia sometimes shows undertones of red. In the United States, an interesting new scenario is now playing itself out: amethyst has recently been discovered in the northern part of the peridot-rich San Carlos Apache Reservation. Locals hope that this discovery will give their economy a much-needed boost. Other sources of amethyst include Sri Lanka, India, Australia, and various sites throughout North America.

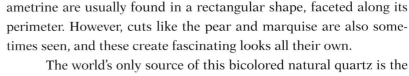

Like most other gems with a long history, amethyst has its share of folklore and superstition. In ancient times, amethyst was thought to quicken the wit, protect soldiers in battle, guard against contagious disease, and bring peace of mind to the wearer. A popular stone with religious leaders, amethyst was believed to control evil thoughts and aid in a person's spiritual development. Early civilizations fashioned amethyst into ornate jewelry pieces and objects of art that included shields, swords, and amulets. Today, Uruguay, itself a minor player in the amethyst market, claims this purple gem as its international gemstone. Domestically, the amethyst is the official stone of the state of South Carolina.

If it were not for its widespread availability, amethyst would be an expensive gem at retail today. While it is true that the deeper colors are the most sought after and generally considered the most valuable, don't let the depth of color be the sole deciding factor when purchasing an amethyst. If a soft shade of lavender appeals to you, for example, don't be swayed by public opinion. When shopping for the darker, more purple variety, take the time to examine your potential purchase from a distance, making certain that it does not look black after you try it on. Although synthetic amethyst has been produced for quite some time, it is seldom encountered at retail, because of the affordable nature of the "real thing." Finally, stay away from dark-colored amethysts in closed settings, for this may indicate a foil-backed, lower-grade stone that is actually too poor in quality even to be classified as a gem.

Ametrine

When amethyst and citrine reside in the same crystal, they unite to form a bicolored gem known as ametrine. This effect, called color zoning, is what makes the ametrine a rare find and an excellent collectible. Because the color-zoning property in this most unusual gem is natural, no two ametrines will ever be exactly alike. In order to properly display this unique effect, bicolored gems like the ametrine are usually found in a rectangular shape, faceted along its perimeter. However, cuts like the pear and marquise are also sometimes seen, and these create fascinating looks all their own.

The world's only source of this bicolored natural quartz is the Anahi Mine, located deep within the hot and hostile jungles of Bolivia. Steeped in folklore and tradition, the mine is rumored to have been first discovered by native Indians in the seventeenth century. Eventually the Bolivian government declared this land a state government reserve and built a fortress near the entrance to the mine. Despite its remote location and the government's military presence, ametrine (as well as natural amethyst and citrine) did make its way across the border into the neighboring state of Mato Grosso, Brazil. These goods were then taken to the city of Corumbá, to be registered and sold on the market as Brazilian. All of this ended with a change in the constitution of Bolivia in 1989. The rights to the mine were sold, and during this past decade, for the first time, these coveted gems began to make their way into the domestic marketplace.

Actually, ametrine was slow to gain acceptance, as suspicious dealers and consumers initially thought it was man-made and either cemented or fused together through some secret process. In fact, the reception was so poor at first that the bicolored rough was split in two, so that the amethyst and citrine could be sold separately. Take it from me, ametrine is a natural gem. I now own a piece of its bicolored rough; it never fails to captivate me under its almost mysterious spell. This controversy was finally put to rest when *Lapidary Journal,* one of the industry's most widely respected trade journals, released an issue that showcased the ametrine on its front cover. Other gem magazines have since featured Bolivian ametrine as well.

Ametrine is among an elite group that I refer to as sole-source gems, because as of now there is just one source of this treasure in the world, which means its future is uncertain at best. I consider any sole-source gem an important find, and ametrine is no exception. If you do not own an ametrine yet, you should add one to your collection as soon as you can. The lovely bicolored ametrine has already begun to carve a niche in the marketplace, fast gaining acceptance by more educated dealers and consumers alike. Unfortunately, this will eventually put pressure on the already unstable sources of ametrine, bringing it closer and closer to possible extinction.

Consumers should know that simulated ametrine has surfaced at retail. Produced in Russia, China, and Japan, it is currently marketed overseas. Logically, it would follow that some lab-created material can be found domestically as well. If you purchase or already own an ametrine and for some reason suspect it is created, simple testing techniques by a trained gemologist can easily distinguish a natural gem from a simulated one. It also helps to deal with reputable sources in whom you have confidence.

Aventurine Quartz

Aventurine quartz is found in a wide array of colors, including green, gold, brown, orange, rust, and blue-green. It is found worldwide, primarily in areas where metals are being mined. It may display a variety of inclusions of pyrite, mica, and other material. As you might imagine, its color pretty much depends on its composition.

In its strongest green form, aventurine quartz can easily be mistaken for jade, as well as for sunstone (also called aventurine feldspar). In fact, its name is from the Italian and literally means "by accident." When aventurine is found with specks of hematite, it produces a dazzling phenomenon known in the jewelry trade as aventurescence. Aventurine quartz is a nontraditional birthstone for August.

Citrine

Citrine is the yellow-to-golden member of the quartz mineral group. Like amethyst, the darker colors are generally more sought after and highly prized than those of a pale pastel hue. A deep golden variety from Madeira, Spain, known as Madeira citrine is a favorite of collectors the world over. In its finest form, Madeira citrine can resemble the rare and costly imperial topaz, which is one reason that citrine is a popular birthstone alternative for those born in November. At one time, citrine was found in the Rocky Mountains of Colorado, though today no American sources are considered to be of any consequence in the overall picture of supply. Other important sources of citrine include Brazil, Bolivia, several countries in Africa, and parts of the former Soviet Union.

Astrologers have linked the citrine to the planet Jupiter, and those born under the sign of Gemini or Leo sometimes claim it as their astral stone. Citrine has been called the "stone of the mind," because of the ancient belief that placing a citrine on the forehead of an elder could bring him great psychic powers. In addition, the citrine has been thought to aid in emotional healing, but only if the stone is found in a shade of light yellow. There may be something to this, when you consider the bright, sunny personality of this wonderful gem!

Another interesting aspect of the citrine is its adaptability in combination with other colors of quartz. The most widely known bicolored quartz is the ametrine; however, few consumers are aware of the interesting smoketrine, a bicolored gem that combines the citrine with smoky quartz. Although the ametrine is indeed natural, the smoketrine is most often produced by heating a single smoky-quartz crystal. On rare occasions it may occur naturally too.

Gem-quality natural citrine is far less abundant than amethyst. In fact, if your citrine shows strong undertones of red, you are probably looking at a heat-treated amethyst. There is nothing wrong with purchasing a heat-treated citrine; as we already know, just about every natural gem in the world is heated to bring out color and clarity anyway. Unless the retailer is well informed, chances are that he or she will not be able to tell you whether it is a heat-treated or natural stone. Unlike the amethyst, large citrines of ten carats or more are not commonly found in conventional retail outlets. As a rule of thumb, then, if the stone under consideration is larger than the state of Utah yet still within your budget, what you are looking at is either heat-treated or synthetic material.

Quartz Crystal

Quartz crystal is a clear variety of quartz that was first found high in the frozen mountains of the Alps. Ancients believed it was formed from ice crystals that would never melt. (In fact, the earliest drinking establishments used quartz crystal in ancient versions of martinis. This worked out just fine, until one of the more macho patrons tried to crack one with his teeth.)

People born in April sometimes lay claim to quartz crystal as their birthstone, while those born under the sign of Pisces, Leo, or Capricorn all claim it as an astral stone. Certain astrologers link quartz crystal with the planet Venus. Like rose quartz, quartz crystal is a very romantic gemstone. In fact, it is widely accepted as the gem for those celebrating their fifteenth wedding anniversary. In addition, like jade, quartz crystal has been said to have cooling powers.

Early soothsayers used this gem to fashion their crystal balls, which are made of man-made materials today. One of the largest natural quartz-crystal balls ever found reportedly weighed over one hundred pounds and was worth hundreds of thousands of dollars. At the other end of the scale, genuine rhinestones, found primarily in the waters of the Rhine River, are sometimes small waterworn pebbles of quartz crystal that have been faceted and backed with a silver foil or metal. Today, quartz crystal is most often seen in bead form when found in jewelry. This variety of quartz is also known as rock crystal, a kind of catchall name attributed to many other forms of crystallized mineral.

At one time, Brazil and Madagascar were considered major sources of quartz crystal. Recent finds have been inconsistent, however, and most dealers now regard them as sporadic at best. Today its most significant sources are the rugged mountainous regions of Switzerland, France, and parts of the former Soviet Union. Curiously enough, quartz crystal is difficult for consumers to find at conventional retail outlets, but this is undoubtedly a result of low demand rather than short supply. Still, because it is so hard to come by, if you do happen across a piece that catches your eye, make it a part of your collection. You'll be glad that you did.

Rose Quartz

Rose quartz is an eye-pleasing pink-to-peach variety of quartz crystal. It gets its color from traces of manganese and titanium within the crystal. Rose quartz is almost always seen in bead form or cut *en cabochon*. In certain rare cases, these rounded cabochons will display the star phenomenon called an asterism. Rose quartz also sometimes contains gold flakes of iron oxide, bringing forth an impressive display of sparkle. Sadly, exposure to prolonged weathering will turn this pleasant pink gem into a rather unattractive shade of slate gray.

Rose quartz is a nontraditional birthstone for January and one of the astral stones of Aries. Like most other varieties of its mineral group, rose quartz is found virtually worldwide. Some of its most important sources are Brazil, Madagascar, and parts of the former Soviet Union. The United States also plays its part in the supply picture: lovely rose quartz can be found in Colorado, Connecticut, New York, and South Dakota.

Rose quartz has been around for thousands of years, with recorded references dating all the way back to approximately 800 B.C. It is a gem that ancients felt promoted gentle healing; its presence was believed to foretell love. In fact, dreaming of rose quartz was suspected to announce the arrival of one's true love. Most rose quartz mined domestically is shipped to Germany and China, where expert carvers turn the pink material into objects of art. The world-famous Field Museum in Chicago numbers among its treasures a stunning hand-carved bowl made entirely of rose quartz. I think this is such an eye-pleasing gem that you should have at least one piece of rose quartz in your gem collection. Its color, I can promise you, will warm your heart.

Smoky Quartz

Smoky quartz is a rather common variety of quartz found in shades of brown, gray, and others in between. Like many other gems, smoky quartz as a rule is heated to bring forth its color. In its more brown variety, it has been confused from time to time with andalusite, as well as a brown variety of tourmaline. In its more gray state, it has been passed off to unsuspecting consumers as smoky topaz, which is more brilliant, harder, and costlier than the quartz variety.

Sometimes found in huge deposits of rough, smoky quartz easily lends itself to giant solitaires in jewelry. I have personally seen solitaires exceeding thirty carats in weight. As you can probably imagine, smoky quartz is found on virtually every continent around the globe, and there is no concern about its future supply. Among its most significant sources are Brazil, Australia, Russia, and Switzerland. The United States has its share of smoky quartz deposits as well. It may surprise you that Arkansas and New Jersey are two of our most important sources. This particular variety of quartz is sometimes noted as an astral stone for those born under the sign of Libra. Although it has never been a major player in the market, Scotland at one time laid claim to the smoky quartz as its international gemstone.

Other Varieties of Quartz

Because it is such a large gem group, the quartz mineral can be found in nearly every color imaginable. Some of these colors are seen in conventional retail venues from time to time, while others may be encountered only at gem and mineral shows. I myself have seen a lovely variety of celadon green quartz and a deep blue quartz that somewhat resembles London blue topaz. Although I'm sure other sources can produce these unusual colors, both of the gems I encountered originated in Brazil. A relatively new find that can come in a variety of pinks and reds (or any color in between) has just started to make its way into the marketplace. Commonly known as strawberry quartz, this unusual variety was first discovered in Asia just a few years back. Because of inclusions, the majority of this material is cut *en cabochon*. An even more recent discovery of a similar material was reportedly found in the Ural Mountains of Russia.

Certain varieties of the quartz mineral group feature inclusions that give these gems a most unusual look. One variety I have seen displays the golden needle-like inclusions known as rutiles. Generally known as rutilated quartz, this is a most interesting and affordable gem. Another quartz gem, known as goldenite, actually contains flakes of golden quartz, which sparkle in the sunlight. Consumers sometimes confuse goldenite with a man-made glass that contains tiny pieces of copper. This material, popular decades back, is called goldstone. That name is misleading: many people believe goldstone is a genuine variety of mineral. To make matters even more complicated, another variety of gem known as aventurine quartz can look very much like goldstone. In fact, it sometimes takes a 10x jewelers loupe to tell them apart. Despite all these similarities, goldstone is strictly man-made glass, and though it is beautiful in its own right, it is not related to quartz or any other variety of mineral.

Various inclusions also can combine with quartz to produce gems that ancient cultures believed to be evil because of their "moving eye." The most common variety is known simply as cat's-eye quartz; it is usually gray to yellow in color, though other colors surface now and then. Always remember to include the word *quartz* when discussing this gem, because some gem purists consider the cat's-eye chrysoberyl the only true type of cat's-eye. Although it is an unusual and attractive gem, cat's-eye quartz is worth far less than the much sought-after color-change chrysoberyl variety. A similar type of quartz, usually brown with black bands and inclusions of asbestos, is known as tiger's-eye. Folklore tells of a link to the regal tiger, and the name was born of such ancient Asian beliefs. The elusive tiger's-eye is one of the stones chosen to represent the astral sign of Gemini. Many consumers have heard of both the cat's-eye and the tiger's-eye; few are aware of yet a third variety, known as hawk's-eye. This seldom-seen quartz is either blue-gray or blue-green, which sometimes causes it to be confused with the cat's-eye when found in a somewhat similar color.

Unlike rare gems of some other mineral groups, these most unusual quartz finds are for the most part inexpensive, which makes them ideal targets for the casual gem buff eager to fill in a collection with stones that are so seldom seen. This doesn't mean you should let down your guard when purchasing quartz. It has been available synthetically for years, and unscrupulous gem dealers will sometimes incorporate up to 50 percent of this material into a batch load, so the consumer really has no option other than to deal with a reputable company that has done its homework. Quality-assurance labs can detect a synthetic quartz from a natural one by running just a few simple tests.

Spinel

References to spinel have long been recorded, yet this mineral group has never really received its fair share of exposure at retail. Unfamiliar for the most part with this lovely gem, consumers have never created much of a demand for spinel. It remains an important group with enormous future potential.

Natural spinel comes in several lovely shades, including red, blue, blue-green, green, and occasionally violet. A rare black spinel that displays a vivid asterism was uncovered in Sri Lanka in the mid-1950s. Today it is a highly sought-after member of the spinel mineral group. In earlier times, a person wore black spinel to signify death and mourning. Consequently, it was considered a gem of great sorrow and deliberately avoided by gem authorities.

In its most vibrant red state, spinel is a dead ringer for the ruby. Compounding this problem is the fact that it is frequently seen associated in nature with rubies, making identification even more difficult. Savvy gem buyers are trained at an early age to select their rubies carefully and to separate them individually from spinels often found in the mix.

Thanks to modern technology, the difference between spinel and the costlier corundum is easily apparent to the expert, but the inexperienced gem buyer has little hope of telling a high-quality red or blue spinel from a ruby or sapphire with the naked eye. Throughout history, these beautiful red and blue specimens were thought to be the finest rubies and sapphires in existence, and they were set in regalia from crowns to swords and amulets. Although large solitaires are difficult to come by, a spinel rough of nearly two hundred carats was once displayed in London during the nineteenth century. Astrologers link red spinel to the sun and blue spinel to the planet Saturn.

The red and blue varieties are among the most sought-after spinels of all. Stones of three carats or more command $300 per carat and up, depending on the quality of the gem and market availability at the time. Since they are costly, red and blue spinels have been produced synthetically for many years. Two layers of synthetic spinel are sometimes fused together by a colored adhesive or other coating, to create a simulant for top-quality tanzanite, emeralds, and certain other select gemstones.

Personally, I believe the market for spinel is undervalued and will change as consumer awareness among amateur gem buffs evolves. Certain varieties of spinel are plentiful, but others are quite rare and could face extinction at any time. These include the violet variety and the black star of Sri Lanka.

Besides Sri Lanka, other important deposits of spinel are located in Myanmar, India, Afghanistan, Brazil, Australia, and parts of the former Soviet Union. Some of these locations produce a wide range of colors, while others seem to specialize in just one or two shades of spinel. In addition, certain European nations contribute to the supply from time to time, and small deposits can be found in various sites scattered about the United States. Should you happen upon a spinel that is appealing and affordable to you, I would grab it now, as consumer awareness is bound to improve sooner or later, causing an increase in demand. As we've all learned in basic economics, when demand increases, prices inevitably follow.

Spodumene

The spodumene mineral group consists of two basic gems: hiddenite and kunzite. Both are found in mountainous regions of the world. Because consumer awareness has been relatively low, hiddenite and kunzite are seldom seen in conventional retail outlets. Both can be found at gem and mineral shows.

Hiddenite

Hiddenite is a yellow-to-green stone that is named after W. E. Hidden, who first described it more than a century ago. It is sometimes found in conjunction with its more famous cousin, kunzite. Chromium and iron are its principal coloring agents, and its most highly prized color is emerald green, which can cause it to be confused with chrome diopside. Hiddenite is one of a small group of gems that will not change color when it is heated. Although it was first discovered in North Carolina, in 1879, and can still be found there, today its most significant source is Brazil. It can also be found in Madagascar and in the state of California. A hiddenite of more than three thousand carats was discovered in Brazil in the mid-1950s, and this remarkable find was later cut into an individual stone that weighs over eighteen hundred carats!

Since the average gem buff is not aware of hiddenite, its market value remains quite low. It is a gem caught in a paradox that others, too, must sometimes evolve from: consumers don't look for it, because it is almost impossible to find anywhere except at rock and mineral shows. Since they never see it, consumers remain generally unaware of hiddenite. Does anyone detect a pattern here?

Kunzite

Kunzite is more popular among gem lovers than its cousin, hiddenite. It gets its lovely pink-to-violet color from traces of manganese. Kunzite was discovered at the Pala Chief Mine near San Diego, California, and named after the noted gemologist G. F. Kunz, who first described it just after the turn of the twentieth century. In fact, that site is considered a secondary source of kunzite even today. At one time Brazil dominated the market, but these deposits are now exhausted. This beautiful gem is sometimes found in Madagascar and Myanmar, but

Afghanistan is currently regarded as the most prolific source. It is most often found in pockets, primarily in mountainous regions, together with hiddenite.

When heated, kunzite will turn green, making it all but indistinguishable from hiddenite. Like certain other gems, it can be restored to its original color through additional exposure to intense levels of heat. Although it is considered a rare find at retail, it is not uncommon to find kunzite in gems of large proportion. Kunzite has an acceptable hardness rating of 7.00 and perfect cleavage (referring to the way a crystal splits along lines of weakness), which actually makes it brittle and most difficult to facet. For this reason, it is regarded as a gem of the evening. In some circumstances, it is possible that kunzite will fade in direct sunlight, when exposed for long periods of time. It can also fade when cleaned in an ultrasonic jewelry cleaner.

Although kunzite is a relative newcomer to the gem world, it has already acquired a history of folklore and tradition, even being considered a modern-day birthstone for February. Those who wear kunzite are believed to be blessed with good fortune. Its soft pastel color is said to stand for purity and innocence, and its presence is sometimes regarded as a symbol of pregnancy and the beginning of a new life. A dream about kunzite is believed to guarantee a hospitable welcome by strangers while on a journey to an unfamiliar land.

Consumer awareness of kunzite is slightly greater than hiddenite, but kunzite is still thought of as an unusual, rare find in the marketplace. Few jewelers offer the gem, once again because of the lack of consumer awareness. This means kunzite remains a gem that is quite rare, yet relatively affordable for all. Unlike some other little-known gems of low awareness, however, kunzite and hiddenite both have limited sources of supply, which someday could cause the market to tighten. Don't read too much into my warnings of fading; with proper care, your kunzite gem will give you many years of enjoyment, for it is truly unique and beautiful. I bought one for my wife a number of years ago, and it has maintained its color even though we live in Florida. So grab one for your very own collection if the opportunity presents itself and it fits nicely into your budget.

Topaz

The topaz is one of the most popular and diverse mineral groups in the world today. Although naturally colored stones are rare, consumers have embraced with open arms the many lovely enhanced colors. Astrologers link topaz in general to the planets Mercury and Venus. Since all varieties can be encountered in one market or another, let's take a look at some of the most popular colors.

The Many Shades of Blue

Without question, the most common and popular color of topaz in the world today is blue. In fact, the thirst for this color is so great that manufacturers have

developed blues within the blues, in shades ranging from a light sky blue through one that some describe as Caribbean blue, to the electric Swiss blue and finally the darkest and costliest form of all, the London blue topaz. Like certain other varieties of topaz, the natural blue gem is seldom seen at retail. In fact, almost all blue topaz seen on the market today began its journey as the white (or colorless) variety. Through a carefully controlled process of heat and irradiation, the stone gradually turns blue. The more the stone is heated, the darker it becomes, until it reaches the deep, almost sapphire-like color of the London blue stone. Since the darker blue stones consume more energy than the lighter blue ones, it follows that the darker kinds also are costlier than the lighter shades. The London blue topaz takes one full year to cool and stabilize, so it is most difficult to anticipate that stone and purchase it in advance.

Because of concerns about rising energy costs in general, I believe that the price of the London blue topaz will someday increase dramatically, or maybe this gem will disappear from the market entirely. I know that experts are probably laughing at this idea, but they do not change my opinion one bit. This has nothing to do with health concerns about irradiated gems, since there has never been one case of related illness or other adverse reaction reported to the Nuclear Regulatory Commission. In fact, the very first jewelry item I ever purchased for my wife was a lovely tennis bracelet, about sixteen carats, of London blue topaz and diamonds; to date, it is still her favorite. Linda also has matching rings, and more than one at that.

Although the dark blue variety is the color of choice for most consumers, each shade of blue topaz has its own look and individual personality. Don't get too hung up on the issue of most expensive or hardest to find, to the point that you ignore the qualities the other varieties have to offer. Although my favorite is the London blue, Linda has all the other shades of blue topaz as well, and she successfully mixes them for some very interesting looks. I recommend that ultimately you have one of each in your collection, if feasible; you will use them all.

As for natural blue topaz, it is found only in the Ural Mountains of Russia, sometimes growing amid deposits of feldspar and smoky quartz. As you can probably guess, this seldom-seen variety is expensive. Fortunately, all of the heat-treated shades of blue topaz, including the London blue, are affordable for most consumers. If the cost of the usual karat-gold setting still limits your choices, seek out the various shades in vermeil or other overlay or in sterling silver. This is an excellent option; all are readily available at retail. Incidentally, all shades of blue topaz are accepted as a birthstone for December.

Champagne Topaz

One of the very few naturally colored varieties of topaz that surface at retail is the champagne topaz. A light-to-medium shade of brown, it is costlier than most heat-treated varieties but far less expensive than most other types of natural topaz. Consumers should find the champagne topaz to be well within their budget. Found primarily in Mexico, the champagne topaz generally forms in areas of volcanic activity. Although not clearly documented, it is suspected to have first been found in an area near the Yucatán Peninsula that is best known for its deposits of the elusive Mexican fire opal. Despite being affordable, the champagne topaz is nonetheless seldom seen at conventional retail outlets. If you see one you like, by all means jump on it quickly.

Imperial Topaz

Imperial topaz is widely considered to be the most sought after of all natural topaz in the world. Its rich golden color is generally not enhanced by heat or any other treatment process. Generally recognized as the traditional birthstone for November, the imperial topaz can prove to be both costly and elusive. An imperial topaz of deep color and red influence can command prices of $500 per carat or more, wholesale. In fact, a seven-carat peach, pink, and raspberry topaz hailing from Brazil was recently offered at prices in excess of $1,000 per carat! This in part explains the popularity of the more affordable and easily found citrine as an acceptable substitute for the November birthstone.

Whenever you come across a reference to "topaz" without a connection to any color, it most likely relates to the imperial topaz. You may come across these references for the months of February, March, and September as well as November. Imperial topaz is considered the traditional astral stone for those born under the sign of Sagittarius, and some astrologers link this gem to the planet Jupiter.

Imperial topaz is sometimes known as sherry topaz or simply golden topaz. If you are in the market for this variety, be aware that impostors sometimes bear similar names in an obvious attempt to dupe the unsuspecting consumer. A commonly encountered stone is the "Madeira topaz," which is actually a golden variety of quartz properly known as Madeira citrine. Certain shades of champagne topaz have also been occasionally passed off as imperial topaz, although a consumer with even a moderate amount of knowledge can tell one from the other at first glance.

Much of the world's supply of this orange-to-golden gem is found in the mining region of Ouro Prêto in the state of Minas Gerais, Brazil. More than thirty thousand carats of cut goods are extracted from this busy area in any given year. The tiny but gem-rich nation of Sri Lanka produces some beautiful specimens as well. Not long ago, in an effort to drive up the price of imperial topaz, the largest mine in Brazil closed for a period of time and worked off existing stockpiles. As it turns out, prices did rise slightly, and top-quality goods may not be available again until well into the next century.

Pink Topaz

The natural pink topaz is one of the most beautiful gems on Earth, but it can be quite costly and is seldom seen at retail. The vast majority of pink topaz seen today is really a yellow variety that has been heat-treated to turn pink. Since most consumers have had little, if any, exposure to pink topaz, the heat-treated variety is easily passed off as the much costlier natural gem. As with other gems that have been treated, the consumer should be able to rely on retailers for this type of information, but as you can imagine, many times they themselves just don't know. As a result, some will guess rather than seem uninformed, and this can be devastating to the unwary consumer. At any rate, surprisingly, even the heat-treated variety is rarely found in the marketplace.

Recently, a new treatment process involving a colored, microthin film has been developed, and some pinks and reds have resulted here as well. This process incorporates a white topaz, rather than a low-grade yellow like the heated variety. This new development is described in greater detail in the next section. Since natural pink topaz can be quite costly, I would put a small, fully refundable deposit on the piece under consideration until the dealer can provide you with a written appraisal or other document that clearly defines the gem as natural. If this is well beyond your reach (as it is mine), all of the enhanced colors and varieties will be excellent options for you. Fortunately, yellow and white topaz both lend themselves to the treatment process very well.

Natural pink topaz, which most gem experts consider the rarest form of topaz in the world, comes mostly from Brazil, Pakistan, and parts of the former Soviet Union. Although some very lovely specimens of pink topaz have turned up in Colorado, the United States will probably never be regarded as even a secondary source of this stunning treasure. Even though it is seldom seen, pink topaz is one of the birthstones accepted for October.

White Topaz

As indicated earlier, nearly all the blue topaz found at retail is actually a heat-treated colorless gem, commonly known in the industry as white topaz. The new method of treatment that involves a microthin layer of film is adding even more demand for white topaz. In this process, a film is applied to a faceted white topaz below the girdle (the widest part of the gem), producing a most unusual effect. Altering the properties of the film produces an almost unlimited number of color displays. Whereas most gems get their color through the conventional method of light absorption, these fascinating creations display their colors through the reflection of light. This coating is thinner than the diameter of one human hair, and it is undetectable to the naked eye. Like all other members of the topaz mineral group, these pieces are durable and can be worn as often as you choose. As with any other gem, of course, exercise common sense to protect the stone from scratching and chipping.

As a result of this evolving technology and consumer demand for blue topaz, white topaz is very difficult to find at retail. Although it is available mostly in small sizes as accent stones, large solitaires can occasionally be located. Since it is relatively easy to facet, white topaz has been seen in a wide array of fancy cuts, producing impressive specimens that are quite exquisite. Like imperial topaz, white topaz is mined from deposits in Ouro Prêto, as well as other areas of Brazil. In 1740, a colossal white topaz was discovered in that part of the country and initially mistaken for a diamond. News spread quickly through the gem community, igniting a frenzy of mining there that lasted for decades. In fact, these rich Brazilian finds were responsible for making topaz the affordable gem it is today. Although much white topaz still comes from Brazil, it can also be found in Sri Lanka, Myanmar, Australia, Japan, several countries in Africa, parts of the former Soviet Union, and a host of other areas worldwide. White topaz may be difficult to find at retail, but labeling it as a rare gem would be inaccurate. Nevertheless, if given the opportunity to purchase a large white-topaz solitaire of several carats or more, you should pounce on it, since it truly may be quite a while before this chance presents itself again.

Other Varieties

Other types of topaz are found from time to time in the jewelry market. Most are heat-treated, often ending up in odd yet beautiful colors that include red, orange, and a multitude of shades of green; one of these displays a lovely aqua-teal color that is a dead ringer for the sea green variety of aquamarine. This color is seldom, if ever, seen at retail, yet for some it is one of the birthstones for August. These colors should not command exorbitant prices at retail, so be certain the asking price is in line with other heat-treated, enhanced varieties you have seen.

The future availability of topaz is of concern only to the gem purist, who will not accept a heat-treated substitute for a natural color under any circumstances. As we have seen, many of the natural colors of topaz are quite scarce and costly, and there are no indications that this situation is going to improve anytime soon. If you decide to hold out for a naturally colored gem, why not just set up your own personal mining expedition to the rugged Urals of Russia? (Of course, it might be best if you first prepare yourself for the climate. To do this, take all the food out of a large freezer, jump in naked, and stay there for about six days. This should acclimate you properly—assuming, of course, that you go in July.) For the rest of us, satisfied to bathe in the beauty of the colorful spectrum of the topaz, future supply is of little concern.

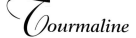ourmaline

Unlike most mineral groups, which are colored by one or two trace elements that reside in the crystal, the tourmaline mineral group is home to a wide variety of these "invaders." This is why the tourmaline family is found in a rainbow of colors considered by experts to span the widest and most colorful gem spectrum of all. Chromium, lithium, magnesium, calcium, vanadium, and titanium (and others ending in *ium* that I can't pronounce) are just some of the elements that are found in the tourmaline crystal and give it rich, vivid colors—which are almost never described as washed out—and a worldwide appeal. Tourmalines are often multicolored, and many varieties are strongly pleochroic, showing different colors when viewed from different angles. Although some of the varied and vivid colors of the tourmaline family are readily available, others are extremely rare and can be very costly. Needless to say, they make the tourmaline mineral group one of the most popular. If you consider the tourmaline your favorite gem, you and I have that much in common, for it is my favorite mineral group as well.

A tourmaline is mentioned as one of the birthstones for May, an astral stone for the sign of Sagittarius, and a stone linked to the planet Mercury. Although it is not clearly identified, the gem in question is most likely a rare, chromium-rich, bright emerald green tourmaline from Brazil that was mistakenly known as Brazilian emerald for many years. (Astrologers link cat's-eye tourmalines, on the other hand, to Ketu, the southern lunar node, where the sun and moon appear to cross paths in the sky.)

In the past, tourmalines were divided into various groups of little familiarity to most consumers, but today the stones are generally classified by color. For example, the stone we know today as the green tourmaline was once called verdelite, no doubt in part to its verd, or green, color. "Green tourmaline" may not sound as scientific, but it serves the purpose.

As with any other mineral group, certain tourmalines will be encountered more frequently at retail than others. Some of these still carry their own individual names to retail, and all are worthy of additional discussion. Overall, they can be considered among the most sought-after gems by consumers worldwide.

Indicolite

Indicolite is the teal-to-blue variety of tourmaline, found mainly in Brazil, Madagascar, Russia, and Sri Lanka. When found in a dark blue state, it is sometimes known as indigolite. As with most other gems, indicolite is almost always heated to brighten and bring out its color, and the end result is nothing short of spectacular. While indicolite in its rich blue state has been confused with fine-quality sapphire, certain other crystals can show an incredible teal color (dark greenish blue) unmatched in nature. Indicolite is sometimes found in spectacu-

lar bicolored varieties, including blue-red and blue-green combinations. On other occasions, indicolite may be found to exhibit a cat's-eye effect; this gem is always cut *en cabochon* to bring forward this special effect. Since I love teal, you can probably guess that the indicolite is one of my favorite gems.

As you can imagine, all of these combinations are next to impossible to find in any venue, as investors grab them sometimes even before they reach their finished state. Indicolite is an especially rare find at retail, with fine specimens costing hundreds or even thousands of dollars per carat in larger sizes. If you share in my passion for the indicolite, be certain to add one to your collection if the opportunity (and the cost) permits, for it may be quite a while before you see another one.

Paraíba Tourmaline

One of the most significant developments in recent gem history was the discovery earlier in this century of blue, blue-green, and violet tourmaline gems of unequal quality in Brazil. Called Paraíba tourmalines after the mine that produced these incredible specimens, they are generally considered the rarest tourmalines on Earth. Demand for the vivid gems could best perhaps be described as a frenzy, and today the flow from the Paraíba mine is approaching an end. When shopping for tourmaline at retail, incidentally, remember that simulated gems of synthetic spinel have surfaced in the marketplace. Although these simulants make beautiful and acceptable substitutes for the highly prized natural Paraíba tourmalines, they should reflect just the slightest fraction of the genuine gemstone's cost.

Rubellite

With its vibrant red color, the rubellite resembles the world's finest ruby, sometimes confusing even the most savvy gem shopper. Fortunately, an experienced jeweler or gemologist can separate the two using simple gem-identification techniques. In certain instances, inclusions transform the clear, sparkling rubellite into a translucent variety, cut *en cabochon* to display a cat's-eye effect. The value of rubellite is pretty much proportionate to the depth of its color; as its red color begins to wane, it becomes pink tourmaline, a popular birthstone for October. Although the pink tourmaline is a much sought-after and beautiful gem in its own right, it is the deep red rubellite that gem purists prize so highly. The rubellite and the indicolite are generally considered two of the rarest members of the tourmaline mineral group.

Some of the most desirable rubellite comes from the Cruzeiro deposit in southern Brazil, although other locations exist as well; the African nation of Mozambique is also an excellent source. Although it is considered a rare find at conventional retail outlets, rubellite can sometimes be found in large weights of five carats or more. In fact, one of the largest deep red gems of all recently surfaced in Brazil. This spectacular gem is said to have weighed in at an impressive twelve carats.

Like the ruby, the rubellite has for a long time been considered a positive, powerful gem. Early astrologers believed the rubellite to be a gift from the sun, making it an excellent treasure for those who are in need of some positive reinforcement in their lives. On rare occasions, rubellite can be found in a bicolored state, usually in combination with indicolite. Although I have been selling gemstones to a nationwide audience since 1985, I cannot recall ever selling even one rubellite of any size or shape. This should say something about its availability and cost. If you stumble on one that you can afford, get it while the getting's good; you may not get another chance for many, many years.

Watermelon Tourmaline

The watermelon tourmaline is a rare gem that displays three different color zones in the same crystal: its natural colors of green and pink are separated by a thin layer of white. The green is said to represent the skin of the watermelon, the slice of white the rind, and the pink its sweet and tasty fruit. It is important to

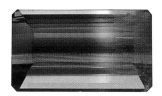

note that the white band separating the green and pink colors is not always visible. Although this variety is still often called watermelon tourmaline, it would be more accurate to refer to it as simply a bicolored tourmaline. Besides this type of watermelon tourmaline, other, less common color combinations also occur in nature, although they are seldom, if ever, available at retail. As in the ametrine, the color-zoning effect in the watermelon tourmaline is 100 percent natural, and like the ametrine and most other gems that possess color-zone properties, the watermelon tourmaline is most often seen as a rectangular step-cut gem. It is found primarily in South Africa, Brazil, Mozambique, Madagascar, and Sri Lanka.

Before I leave this subject, I would be remiss if I didn't mention that the watermelon tourmaline is my youngest son's favorite gem, even though I told him not to eat it after the fireworks on the Fourth of July. He, too, is developing a penchant for gems and jewelry, and last Christmas I purchased a loose watermelon tourmaline for him, which he totally enjoys. Like father like son, I guess.

Other Colors

Although most people consider the green and yellow varieties of tourmaline the most common, its range of colors is almost endless. It is important to note here, however, that the green tourmaline we are speaking of is one that is best described as bottle green in color. The other chromium-rich green tourmaline, which was discussed earlier as an astral stone and birthstone, is very rare and quite costly. Other names you may encounter on the market for this particular species of tourmaline are Brazilian tourmaline and chrome tourmaline.

The pink tourmaline is probably the most popular member of this colorful gem group. This lovely gem can be a deep bold pink, or a soft pastel version that may be reminiscent of the lovely pink sapphire. If you were born in October, the pink tourmaline is one of your birthstones. Pink tourmaline is sometimes heated to a colorless state, though this variety also occurs naturally.

Tourmalines may also be found in varying shades of brown, gold, and even as a mysterious black stone sometimes referred to as schorl. This black variety can also be obtained by exposing a rubellite to extreme heat; however, because of the value of this ruby red tourmaline, this is a practice that is almost never done. Suffice it to say that heating a rubellite to yield a schorl would be like feeding a French soufflé to a goat.

The geographical world of the tourmaline is almost as widespread as its color spectrum. Among its sources, Brazil is probably the most important. However, fine-quality tourmalines can also be found in Sri Lanka, Myanmar, Madagascar and elsewhere in Africa, and parts of the former Soviet Union. In the United States, Maine and California also contribute some colorful specimens.

Because of its many family members, the future supply of the tourmaline mineral group is all but impossible to predict. As of now, certain varieties, like the one from the world-famous Paraíba mine, will someday be extinct, while others, such as the bottle green and yellow varieties, look to be in good supply for many years to come. To be certain, however, whenever you do come across a beautiful tourmaline variety of an uncommon color, it would be a very wise move to add it to your collection immediately. You'll never really know when the chance will come around again.

Turquoise

One of the oldest and most popular of all gems is the turquoise. A stone of historical and even biblical significance, turquoise was mined by the Egyptians in the Sinai Peninsula. Still other ancient cultures, including the Greeks, Romans, Incas, and Mayans, were all enthralled by its unparalleled color. In the Victorian era, children often wore turquoise inset in gold lockets and rings. According to legend, the first person ever to wear turquoise was Isaac, the son of Abraham.

The rich blue color of turquoise depends on the ratio of copper and iron found in the content of the rough; copper contributes to the blue part of the spectrum, while the presence of iron turns the blue to green. The crystal also is often invaded by manganese oxides, which contribute a variety of black lines sometimes known as veins. These veins contribute to the character of turquoise; because of this, no two stones will ever be exactly alike. In addition to being cut and polished *en cabochon*, turquoise is also seen fashioned into beads or nuggets. Because it is sensitive to heat, turquoise is a most difficult gem to polish, and the craftsperson must make certain not to allow the stone to overheat through friction in the polishing process, or the stone will lose its attractive blue color.

Turquoise is generally accepted to be an opaque gem, but on occasion it may be found in crystal form. The first turquoise crystal on record was discovered in Virginia in 1912. As you can probably imagine from its blue-to-green color range, clear turquoise crystals are indeed quite beautiful, rare, and expensive.

When shopping for turquoise, always remember that some specimens are very porous and may crack or even split completely over time. To overcome this problem, a resin or waxlike substance is used to help maintain the integrity of the gem's structure. Generally speaking, the most sought-after color of turquoise is best described as "robin's-egg blue." From time to time you may also encounter what is called a reconstructed or reconstituted turquoise, which in fact is small bits of the natural gem artificially fused together. In addition, be aware that a lab-grown variety has been on the market for about twenty-five years. Turquoise is most often seen set in sterling silver, either alone or in combination with other gems cut *en cabochon,* such as malachite, onyx, and carnelian. Although most of the turquoise jewelry I have seen is associated with designs from the American Southwest, pieces with an antique look can sometimes be found in association with hematite or marcasite.

Turquoise is the traditional birthstone for December, but those born in June or July also claim it as a birthstone. Some astrologers link turquoise to the planet Venus and the astral signs of Aquarius and Capricorn. Earlier cultures believed turquoise had the power to protect people from the bite of a poisonous snake, and it was said to be able to overcome sight disorders and even blindness.

Turquoise is often found where copper deposits exist, many of them in the blistering deserts of the world. Many people regard the pure light blue stones from Iran as the world's finest, while the United States contributes a variety that tends to run a little more toward green with a little less blue. Domestically, New Mexico, Arizona, Nevada, California, Colorado, and Utah all contain important deposits of turquoise. Internationally, Mexico, China, Australia, and parts of Africa also are major players in the market. The original sites linked with ancient Egypt and the Sinai have long since been depleted, and the relentless desert heat sometimes leads to inconsistency of supply. Although small, pure blue crystals without inclusion are rare, the turquoise market in general is pretty much regarded as stable and secure.

Zircon

The oldest-known rock in the world is generally believed to be a zircon from Canada estimated at nearly 4 billion years old, a pretty significant statement when you consider that scientists estimate the Earth to be about 4.6 billion years of age! A gem of excellent density and sparkle, the zircon rates highest among all gem-quality stones (excluding the borderline gem hematite) on the specific-gravity scale, and only the diamond has a higher refractive index. Because of these properties, colorless zircon—devoid of impurities—is a dead ringer for the diamond to the untrained eye. (It is also a very distant relative of the cubic zirconia.) Although not so common these days, colorless zircon was marketed for years as a diamond substitute to those who would not settle for a man-made replacement. As you can probably imagine, some consumers have been taken in by the colorless zircon, believing they have purchased a high-quality diamond while enjoying a substantial savings.

Zircon is also found in a wide array of vivid colors, resulting from different impurities within the stone. Most popular among consumers is the blue variety, which some people regard as a birthstone for September and December, and which is considered an astral stone for the signs of Aquarius and Sagittarius. This lovely blue gem begins its life as a rather unattractive dark brown stone, found primarily in Thailand. Heat and pressure transform it into the blue or colorless variety. Through certain modifications in the enhancement process, these brown gems can also be heated to a sparkling yellow hue. Interestingly, if the blue gem reverts to brown as a result of exposure to the elements, it can be returned to its blue color by subjecting it to additional heat treatments.

Besides the ever popular bright blue, the zircon's colors include pink, green, red, orange, yellow, and a honey brown variety often known in the gem trade as hyacinth (see page 119). Unfortunately, most of these natural colors are seldom, if ever, seen at retail. At times, the zircon also has been known to display various gem phenomena, including aventurescence, chatoyancy, and color play. Because of its storied past, the zircon is deeply rooted in folklore and tradition. As a group, it is recognized as a birthstone for April and August; pink zircon is also considered a birthstone for October. Astrologers link the zircon in general to the planet Venus, and the rare red-orange zircon to Rahu, the northern lunar node, where the sun and moon appear to cross paths in the sky. Various other varieties and colors have been said to offer protection from lightning, intense heat, grief, and sadness. People have also believed that the zircon can ward off evil spirits and nightmares.

The zircon is not without drawbacks. It is a brittle gem with a strong tendency to chip, so it should not be worn during many types of activities. Because of its rather fragile nature, the zircon is considered among gem cutters to be one of the most difficult stones to facet. In addition, zircon contains low levels of radioactive elements that eventually cause the gem's crystal structure to break down. This may result in a loss of color and luster (or inner glow), but the amount of radioactivity present poses no health threat to the consumer. Such stones are commonly known around the gem trade as low zircons, a name that makes them about as popular as an ice storm in March.

If all this doesn't deter you, I wish you luck shopping for zircon at retail, because, believe me, you're probably going to need it. Still, if you persist long enough, your efforts may prove worthwhile. If you do discover a retailer that carries a selection of zircons, resist those gems that are set high in pronged mountings and be practical when making your choice. (On the other hand, my wife has a gem with a prong setting, and she has never had a problem yet.) In general, look for bezel or channel settings that offer some protection; stay away from bracelets, as they take much more abuse than other kinds of jewelry. In all honesty, zircon is most suitable for earrings and pendants, although gems in well-protected settings will also work nicely in rings. If you want a specific zircon in a suitable setting, be prepared to go the costly special-order route. I know this firsthand: it's the only way I could get one for my wife. Since zircon is also produced synthetically, get some sort of assurance in writing from the dealer as to its authenticity: what a shame it would be after all that searching to end up with a lab-created substitute you didn't want in the first place!

Zircon is a rare find in the jewelry marketplace because of slow acceptance by consumers, rather than a lack of supply. As long as demand is reasonable, overall zircon will remain available, in no imminent danger of extinction. Still, there is a pale blue variety, sometimes called a starlite, that experts consider a rare gem. Zirconite, a pale yellow member of the mineral group, is also nearly impossible to find, though this variety is not as widely known nor as popular with serious gem collectors as its light blue counterpart.

The most important source of zircon is Sri Lanka, where it occurs in gem gravel along with other lovely finds. Other sources include Myanmar, Thailand, Cambodia, Vietnam, Australia, parts of Africa, and Brazil. It is ironic that zircon can be found virtually worldwide, yet the gem is almost impossible to find at retail.

Hyacinth

About the only variety of zircon you are liable to encounter at retail is the honey brown color commonly known as hyacinth. This gem is well known throughout the jewelry world, but it has never really caught on with modern consumers. Every so often, when earth tones rule the fashion market, the hyacinth temporarily comes back into style, yet it has never been able to establish a place of importance. Even among zircons, the hyacinth lags behind the blue and colorless varieties of the mineral group. Ironically, heating a hyacinth will turn the gem colorless or blue.

Although hyacinth is a minor player today, this was not always the case. In fact, it was once so popular that for a while it seemed to become extinct. Because of its earlier popularity, the hyacinth has more than its share of folklore and tradition specifically associated with it. People born in January, February, or March sometimes consider hyacinth their birthstone, while those born under the sign of Leo claim it as their astral stone. In the Orient, the hyacinth was linked to the dragon that was thought to control the solar and lunar eclipse. It was also often associated with Jupiter and Mars, as well as the sun and moon. Throughout history, hyacinth was credited with reducing the pain of childbirth, protecting against evil spirits, aiding in weight loss, and improving physical strength. Many early explorers took the hyacinth with them on their journey, for it was believed to protect them from contagious disease and injury. Some scholars in the sixteenth century carried this gem with them and placed it on their forehead to induce sleep when their mind became overtaxed.

Although it is easily the most common of all zircons, the hyacinth is nonetheless seldom seen at conventional retail. If you do happen across one that is attractive and within your budget, you should make it a part of your collection. Remember, too, if looking for hyacinth, do not confuse it with gems that have a somewhat similar outward appearance, such as hessonite and Madeira citrine.

Zoisite

Little was known of the zoisite mineral group before tanzanite swept the gem world by storm in 1967. Although that blue gem is obviously the most important variety, the zoisite group does have other members and other colors, including the pink-to-red variety known as thulite.

Tanzanite

In a gem world in which stones date back thousands and even millions of years, tanzanite is just an infant. First discovered in 1967 in a single deposit in Tanzania, tanzanite was initially believed to be a form of sapphire that would rival the Ceylon for quality and color. Probably the trendiest gem on the market today, tanzanite will soon eclipse even the ever popular (and more affordable) amethyst in retail sales, according to some experts. Some astrologers link tanzanite to the planet Saturn.

Out of necessity, nearly all tanzanite found at retail today is heat-treated at low levels and then cut to show a deeper blue. As I previously mentioned, this is commonplace in the colored-gem world, and so it does little to affect the gem's market value. Tanzanite is a stone with strong pleochroism, generally appearing not only blue but also purple and slate gray, depending on the angle from which it is viewed.

Because tanzanite is a sole-source gemstone, found only in one country, its future remains in doubt. Goods that once flowed freely into the marketplace have been reduced to a trickle, thanks to devastating floods attributed to the weather phenomenon known as El Niño. Thus, a cloud of uncertainty hovers overhead, and the search for new sources is intense. Complicating matters is the fact that even if there are other tanzanite deposits, the rough is located far beneath the surface of the Earth, too far for miners to reach at all. A large infusion of capital is needed to further explore these areas, and investors are understandably hesitant to back these efforts with millions of dollars that may—or may not—result in any

return whatsoever. Mining for tanzanite is truly a hit-or-miss type of situation, which has turned some hopeful miners into millionaires while forcing others into bankruptcy. One solitary miner became so wealthy from his tanzanite good fortune that he once climbed to the roof of a hotel and showered his fellow miners, waiting below, with thousands upon thousands of dollars!

Needless to say, this is definitely a sellers' market. The demand from consumers is high, while supplies are at best uncertain. So just how much tanzanite is left in the world? Well, it depends on whom you talk to. Some predict its demise in less than ten years, while others say we won't have to worry about it in our lifetime. Personally, considering these recent developments, I think its future is in doubt even today. There is one thing, however, that all gem dealers do agree on: once its supplies are depleted, tanzanite is likely to disappear forever. For now, gem dealers pay locals to watch for new finds, and they will often place orders sight unseen for the gem rough; future parcels are auctioned off, usually one year in advance. Obviously this practice will most likely change in the near future.

Although they are quite costly, large solitaires of three to five carats are available at high-end jewelry stores, through gem dealers, and by special order from almost any broker with a fax and a line of credit. For well-heeled investors, a large tanzanite solitaire would be a very worthwhile addition to your collection. Be prepared to plunk down some major dollars if you want nothing but the best, because top-grade solitaires of considerable size will cost thousands, if not tens of thousands, of dollars. For the rest of us (notice I said *us*), affordable bands and clusters can be readily found at retail.

The high cost of large tanzanite solitaires has not been lost on the industry, and a number of simulants have already developed, while others are being perfected as of this writing. A new variety recently introduced to the consumer marketplace is a close relative of yttrium aluminum garnet, a diamond simulant more commonly known as YAG. According to some accounts, this tanzanite look-alike is said to be the closest in appearance yet to the real thing. In addition to this newest entry, consumers searching for alternatives to the expensive tanzanite solitaire can choose from simulants of synthetic quartz, corundum, and spinel.

If your wardrobe does not include a tanzanite, definitely add one (or more) to your collection as the opportunity presents itself. As with any rare gem, try to find the largest individual solitaire that your budget will allow, but do not sacrifice quality for size. Whether shopping for a solitaire, a cluster, or a band, look for those stones that are vivid in color and range from blue to violet. It is impractical to expect to find a $5,000 tanzanite on a budget of $200, so set realistic goals that fall within your means. Remember: any shade of natural tanzanite will be better than owning none at all.

Thulite and Other Varieties

Thulite, a pink-to-red variety, is without a doubt the second most important member of the zoisite group. Although its predominant color is pink, some stones take on soft tones of orange. Thulite may be found as a sparkling crystal, though more often than not it is opaque, usually with undertones of green, gray, or white. Records dating back to its earliest finds in Norway tell of its brief popularity in jewelry years ago. Besides Norway, thulite has been found in Austria, Namibia, Australia, and in the gem-rich state of North Carolina. Thulite may be the second most common form of zoisite, but it still remains nearly impossible to come by, even at gem and mineral shows. I myself have never seen one anywhere.

Besides thulite, there are other varieties in the zoisite mineral group. Their colors include gray, white, yellow, green, and brown. They are generally just named in association with zoisite, meaning the gray variety is called gray zoisite, the white one is white zoisite, and so on. I mention them here so that you know they exist, but I wouldn't worry about trying to find one; your chances are remote at best.

Chapter 10

Other Gems and Minerals

Andalusite

Sometimes known as "the poor man's alexandrite," andalusite does in fact resemble certain alexandrites at first glance, but that is where the similarity ends. Andalusite does not contain the color-change properties of alexandrite, and its cost is a mere fraction of the market price for alexandrite. Still, it is an interesting stone, known for its pleochroism, showing color combinations of gold, brown, and green when viewed from different angles. It is a gem that is normally found in alluvial deposits of pebble size. I have never seen even one gem of more than 1.5 carats in weight. Andalusite was named after Andalusia, the region in Spain where it was once found in abundance; today, the primary sources of andalusite are Sri Lanka, where it displays a deep green undertone, and Brazil, where stones are generally lighter in color. At times, andalusite from Australia also surfaces in the marketplace.

Besides its use in jewelry, andalusite is employed in the making of certain types of porcelain, as well as optical products. Chiastolite, a variety of andalusite, was feared by early cultures and called "the stone of death." This is because a black cross is easily visible in the depths of the stone, as a result of its inclusions. Chiastolite is seldom seen anywhere at any retail level, and in light of its reputation, that is not necessarily a bad thing.

Apatite

Apatite in its blue-green state closely resembles the costly Paraíba tourmaline from Brazil; other than color, however, the two have nothing in common. Apatite is readily affordable, while the Paraíba tourmaline is considered within reach only by men who own oil fields in Kuwait. Still, it is a beautiful gem that deserves some recognition for its appeal to the eye. Blue apatite is considered a nontraditional birthstone for December, and astrologers often link it to Ketu, the southern lunar node, where the sun and moon appear to cross paths in the sky. It hails from any number of places throughout the world, including Myanmar, Sri Lanka, Brazil, parts of East Africa, Mexico, and Spain.

While the turquoise variety is the color most often seen at retail (when it is seen at all), other colors can sometimes be spotted at gem and mineral shows. The most sought-after color of all is violet, but it also is known to occur in lovely shades of pink, yellow, and a green from Spain known in the trade as "asparagus stone." Occasionally it is also found in a colorless state. Blue and green varieties are dichroic, showing a second color when viewed from a different angle. At times, apatite displays a stunning cat's-eye effect. About the only thing that keeps apatite from enjoying a pivotal spot in the marketplace is its hardness, which is only 5.00. Because of this, it is better suited for pendants and earrings than for rings or bracelets. Still, the color of apatite is beautiful, making it hard in one way: for collectors to resist it.

Azurite

Azurite is a beautiful blue mineral most often found where copper deposits or large limestone pits are being worked. It is similar in structure to malachite; in fact, when they combine, the result is a seldom-seen gem known as azurmalachite. Azurite comes from Namibia, Australia, parts of the former Soviet Union, China, and the southwestern United States. At one time, France was the most significant source of azurite, but these deposits are now exhausted.

Nearly all azurite occurs in a translucent form; crystals are seldom encountered. Its hardness is only 3.50, and it is quite difficult to cut properly. Even today, azurite is also pulverized into a powder and used as a pigment for the paint industry; this tradition dates back to earlier cultures in the Far East. Azurite is sometimes mentioned as an astral sign for Taurus. If you find one, grab it, but handle with care.

Benitoite

The little-known gem called benitoite first surfaced just after the turn of the twentieth century around San Benito Mountain in California. Although benitoite has turned up in other locations throughout the world, to this day San Benito County has remained the world's only source of gem-grade material. As a result, benitoite was recently declared the official state gemstone of California.

Two of the elements influencing the color of benitoite are titanium and barium. It generally spans the range of blues; once in a very great while, pink and even colorless benitoite have been recorded. The blues generally rival fine-quality sapphires and tanzanites, making this one of the most eye-catching gems of all. The blues are also strongly dichroic, appearing colorless when viewed from a different angle. Although the pinks are natural, the colorless variety is generally produced by heating a blue gem that has lifeless color.

Benitoite has a moderate hardness of 6.00–6.50, and it is a fairly dense gem, with a specific gravity of about 3.65. Since it is a somewhat brittle stone, benitoite must be handled with care. I wouldn't plan on seeing one of these too soon: a total of just four thousand carats of gem-quality benitoite has been produced in nearly one hundred years!

Calcite

Calcite is one of the most common minerals found on Earth. It often forms within the veins of huge slabs of limestone. Although it is found in a wide variety of sources, the marble-rich area around Franklin, New Jersey, is home to many minerals found residing within calcite deposits. Located near the New York border, Franklin has a most interesting museum that showcases calcite glowing in a multitude of colors under fluorescent lighting. This fluorescence is present only in calcites that also contain impurities of manganese. Geologists have found evidence of yellow calcite dating back millions of years in what is now the state of Utah. Yellow calcite is sometimes found in combination with brown aragonite and other minerals; this is known as septarian. A colorless variety of calcite is used in the production of fine-quality Italian marble. In fact, this same colorless variety is sometimes seen cut *en cabochon* to display a cat's-eye effect. Calcite can also be fashioned into beads and various artifacts. A brown-and-white-banded stone, often used in sculptures, that is known as Mexican onyx is not onyx at all but a variety of calcite. Calcite is a fascinating mineral that often displays an interesting optical illusion caused by double refraction: when a person looks through the crystal, it splits the light rays in two, producing a double image. If possible, pay a visit to Franklin sometime; the museum alone is worth the trip.

Danburite

Despite rumors currently running wild throughout the gem world, danburite was not named for me at all. In fact, it was named after the place where it was first discovered: the town of Danbury, Connecticut. Today, however, most danburite comes from Myanmar, Japan, Mexico, and Madagascar. If you live near Danbury, I wouldn't bother with the pick and shovel: this area now yields less than gem-grade material. The majority of danburite seen at retail is colorless, although occasionally it may also have a light pink or yellow undertone. In its colorless state, it is so brilliant that it can be easily confused with a white sapphire or a diamond. Because of this, danburite is sometimes seen in closed settings with clear or colored foils, an enhancement known as foiling. There is nothing inherently wrong with purchasing a foil-backed gemstone, as long as the seller clearly identifies it as such to the consumer.

Diopside

Diopside is a distant relative of kunzite and hiddenite, two varieties of spodumene discussed earlier. It can be found in a multitude of colors, but the one seen most often is a bright emerald green, in a gem known as chrome diopside. Besides chromium, the colors of diopside depend on the content of iron and magnesium; as the iron content increases, the stone becomes darker. All the colors of this gem that I have seen are vivid rather than washed out. In ancient times, some people believed green diopside had fallen from the tree of life, and therefore the dead should be buried with a diopside to guarantee renewal of life. Green diopside is also associated with peace and tranquillity. In fact, in certain cultures people placed a diopside on the forehead prior to resting. This was done in order to ensure sweet dreams while sleeping.

Diopside comes primarily from Myanmar, India, Brazil, parts of Africa, Russia, and the United States. A lovely violet-blue form of diopside, seen mainly in bead form at retail, hails from New York State. A new variety of diopside has surfaced in the state of Tamil Nadu, on the southern tip of India. It most closely resembles the green tourmaline and often displays an asterism. The gems from India were found in association with a deposit of emeralds. When it comes to diopside, by all means make one or more part of your gem collection, but beware of one thing: it rates only a 5.50 on the Mohs' hardness scale.

Dioptase

The first recorded finds of dioptase date back to 1780, when it was discovered in Russia. Today, other sources include Congo, Namibia, Chile, and the United States. Dioptase was believed to be a variety of emerald until advances in gem technology disclosed its true identity. It was reclassified just after the turn of the nineteenth century. From that moment on, gem experts treated dioptase with little or no regard. In fact, even today it is often called "the emerald of the poor." Dioptase is commonly found in association with copper ore, even in combination with malachite. For this reason, dioptase was also incorrectly described as "the copper emerald." As you might guess, the southwestern United States (in particular, the state of Arizona) is the most important domestic source of dioptase. Because of its low hardness rating (5.00), dioptase is seldom seen in jewelry at retail. If you're at all curious about this lovely stone, try your luck at a local rock shop or a gem and mineral show. Although dioptase is considered a rare mineral, it does not command a high price in the marketplace. This is probably a result of low consumer awareness and demand.

luorite

Fluorite was first discovered in 1530 in England, an important source of supply to this very day. It can also be found in Mexico, Norway, Russia, and China. In the United States, deposits exist in Illinois, New Hampshire, New Jersey, and New York State. Fluorite, which was once also called fluorspar, is essentially a mixture of calcium and fluoride. Like calcite, fluorite is often found within slabs of limestone. Besides its use in jewelry and crystal artifacts, fluorite also plays a role in the manufacture of steel.

Fluorite—with a little helping hand from humans—has some interesting properties. It can be seen in a wide array of colors, often more than one simultaneously in the same crystal, much like the ametrine quartz. This is known as color zoning. When heated, some fluorite glows in the dark. In addition, certain crystals change color in the presence of fluorescent lighting. This phenomenon is known as fluorescence.

Fluorite may be found in shades of lavender, green, blue-violet, yellow, brown, pink, and black, among others. It also often appears in a bicolored state, combining various interesting looks into one special stone. The stone can be heated to produce this bicolored effect, but this is a tedious and costly process that is seldom used. Therefore, nearly all bicolored fluorite found at retail is natural.

Because of fluorite's low hardness rating (4.00) and its tendency to chip, some experts classify it as a mineral rather than a gemstone. Despite this, fluorite is currently enjoying newfound popularity as a trendy gem of the 1990s. Many people believe it has a calming effect on the body, and because of this it is sometimes used in the art of massage. During the eighteenth century, it was ground into a powder and mixed with water to treat kidney disease. Regardless of any medicinal value, fluorite is an intriguing gem that's well worth a trip to the local rock shop or a gem and mineral show when it comes to your town.

Hematite

Hematite is a variety of iron oxide that is usually found in a silver or black state. On occasion it can also display shades of deep red. Hematite is sometimes fashioned into costume jewelry, in combination with black onyx or various other chalcedonies. In fact, hematite is believed to be among the first minerals ever fashioned into jewelry. It may be seen cut *en cabochon* or brilliantly faceted on its table (top). Besides jewelry, hematite is used in the carving of artifacts. It is also often ground into powder for use in polishing other gems and minerals or as a base for pigments. In fact, Native Americans believed that war paint made from this mineral would make them invincible in battle. Other people, in the eighteenth and nineteenth centuries, wore hematite during mourning.

The most important source of hematite is Brazil; other places where it is found include China, Canada, England, and Germany. Hematite is a nontraditional birthstone for March and December. It has deep-seated meaning to ancient astrologers, who associated this glittering mineral with Mars and Mercury. Those born under the sign of Aquarius often claim hematite as their astral stone. Some earlier cultures believed that hematite could stop bleeding. It sparkles, it's cheap, so what the heck? Go for it!

Iolite

Iolite is considered the only gem-grade variety of cordierite found in the world today. Most experts consider iolite a gem with very strong pleochroic properties, meaning it may show many colors from different angles. The predominant color is blue, with flashes of violet, gray, or yellow, or all of these.

Iolite is one of those gems that walks a very fine line when it comes to classification. Generally, the transparent blue crystals found in Sri Lanka and Madagascar are the only varieties of iolite that experts recognize as being of gem quality. Obviously this makes those countries the two most significant sources of iolite today. These blue gems are found in alluvial deposits, primarily in the form of water-worn pebbles; this is why iolite is sometimes known as "the water sapphire." Since such pebbles are quite small, large solitaires are virtually nonexistent. In fact, in the twelve years I have sold gemstones, the largest gem-grade iolite solitaire I can recall was just 1.1 carats in weight! Iolite can also be found in lumps that are embedded in granite and other rock formations, or in lava rock where volcanic activity is common.

Iolite is a fascinating stone with a storied past. Ancient mariners used a blue-violet crystal as a compass to guide their way when out to sea. It was noted that this crystal would show different colors when held to the northern sky than it would when held to the southern sky. Because of its vivid pleochroism, most gem historians now believe that this crystal was in fact the iolite. Folklore tells us that prior civilizations believed iolite was the key to unlocking creativity in an artist. Today, astrologers often link iolite to the planet Saturn.

Iolite is not generally found in conventional retail outlets, but rather at gem and mineral shows. This is more a result of low consumer awareness than rarity. Gem cutters consider iolite a difficult gem to work with, because it can lose its attractive pleochroic properties if it is not cut properly and its crystal structure will often chip during the carving process. Besides jewelry, iolite is used in the production of certain optical coatings. Iolite is a lovely blue-violet gem that should be part of your collection.

Kyanite

Kyanite is a most interesting gem. Its hardness rating (5.00 and 7.00) actually varies with the direction of its cleavage. Kyanite is found in mountainous regions, sometimes associated with andalusite and corundum. In the United States, deposits of kyanite have been discovered in Montana and North Carolina. It can also be found in Switzerland, Brazil, and Myanmar.

Originally called disthene, kyanite dates back to discoveries during the nineteenth century. The most highly prized stones are those with a vibrant blue or blue-green color. Kyanite also comes in shades of gray, yellow, and white, as well as a colorless variety that is considered quite rare. Because of its poor cleavage and brittle nature, natural kyanite is seldom seen in jewelry. However, increasing consumer demand has resulted in a wide variety of simulants that are both hard and durable. These look-alikes are generally made of synthetic corundum, quartz, or spinel. Most simulated kyanite found at retail today displays a vivid blue hue.

It was once believed that a kyanite suspended from a human hair could follow the Earth's magnetic force like the needle of a compass. Some earlier travelers took kyanite along with them when entering unfamiliar territory. About the only place today where you will see natural kyanite is in rock shops or at gem and mineral shows.

Malachite

Malachite is a banded green stone found associated with deposits of copper, often together with a blue gem known as azurite. In fact, malachite and azurite sometimes combine in the same stone to produce incredible bands of blue, black, and green. This colorful gem is commonly known as azurmalachite. Although both malachite and azurite can be found in abundance, azurmalachite is seldom seen at retail. An important find of malachite was uncovered in the Ural Mountains of Russia during the eighteenth century. Once considered the world's most reliable source, these formerly extensive deposits are now pretty much a thing of the past. The Russians treasured malachite so much that they were the first to successfully produce a synthetic alternative. Today, Namibia and Congo are the two most reliable sources of malachite. Sources in the United States include Arizona, New Mexico, and Oklahoma.

Malachite has been around since ancient times; consequently, it is a gem rich in folklore and tradition. Significant to astrologers, malachite is linked to three astral signs: Aries, Scorpio, and Sagittarius. In its heyday, Russian nobility wore malachite as a statement of their wealth. It was once used in columns and as a decorative inlay for some of the more prominent buildings in Moscow and other Russian cities. In fact, some evidence of this costly architecture still exists. Earlier cultures believed that malachite reduced the labor of childbirth. It was also often placed around the neck of a baby to arrest the pain of teething. In addition, malachite was once pulverized into a powder form and used in the making of eye shadow.

In jewelry, malachite is most commonly seen set in sterling silver, usually in settings influenced by the American Southwest. Less commonly, it can be found as jewelry from the Victorian era; however, these are often reproductions rather than originals. Thanks to Mother Nature, no two stones will be exactly alike, and this makes matching stones nearly impossible. Because of this, malachite is most often seen in singular settings, such as rings and pendants. It is a most interesting and affordable gem, which you should add to your gem collection when the opportunity presents itself.

Milarite

Milarite is a little-known mineral that is found in shades of yellow and yellow-green. Initially, miners mistook milarite for beryl and hacked away the yellow from the green, leaving useless scraps of milarite in their wake. A relatively new stone, milarite first surfaced in Namibia in 1962. A second deposit was found in Mexico, by miners prospecting for feldspar and quartz. Recent finds high in the Swiss Alps have raised hopes that additional deposits may be found in other rich mountainous areas, such as the Urals, the Himalayas, or even the Rocky Mountains of the United States.

Serious gem collectors quickly consumed the very few pieces of gem-grade material that were discovered. None of these earlier finds have ever made it to retail. Unless there is a major discovery in the future, there is little chance that the average consumer will come across milarite in the marketplace. If you do, however, handle it with care: milarite rates just 5.50 on the Mohs' hardness scale.

Moldavite

Moldavite is the bottle green variety of the tektite mineral group. It is a natural glass, similar in structure to obsidian. The only known source of this extremely unusual gem is the Moldau River valley in the Czech Republic. As you might suspect, this is how moldavite got its name.

Although the first evidence of moldavite dates back to the eighteenth century, it is believed to be one of the oldest gems on Earth, perhaps going back to the beginning of time. At first, moldavite was believed to be a result of volcanic activity. Today, however, there are two theories about its origin: some geologists believe it was actually formed within the Earth as a direct result of heat and pressure from large meteorites that were embedded in the planet; a second group theorizes that moldavite is a part of the meteorite itself that formed after cooling in Earth's atmosphere. For this reason, moldavite is sometimes called "the stone of the heavens."

Few people know that the tektite group has another member, a brown-and-green variety found only in Thailand. This is sometimes seen in the production of vases and artifacts; it is not considered gem material. It is unlikely you will come across this second variety, and it sounds to me as if you're not missing much.

Obsidian

Obsidian is formed as lava from volcanic eruptions cools within the Earth. A natural glass, it has a composition similar to that of moldavite. Obsidian can be found anywhere volcanic activity has been documented. Some of its most important sources today are South America, Japan, Mexico, Afghanistan, and parts of the United States. In fact, Arizona, Idaho, Montana, New Mexico, the state of Washington, and Hawaii all contain deposits of obsidian.

Although its most common color is black, obsidian also can be found in various shades of gray, brown, and deep blue. Different locations have contributed other unusual colors to the group; for example, a recent find in a mountainous region of Afghanistan shows varying shades of teal, mint, and sky blue, while Mexico contributes an unusual multicolored variety called rainbow obsidian. The San Carlos Apache Reservation in Arizona, for years the most significant source of peridot, also produces an interesting banded variety of obsidian known locally as "the tears of the Apache." In fact, early Native Americans used obsidian for the sharp tip of an arrow or spear, creating objects for hunting and war.

I would be remiss if I moved on without mentioning a product known as "Mount Saint Helens stone." Although it does actually contain components from the 1980 eruption of Mount Saint Helens in the state of Washington, this product is essentially a form of man-made glass. Since nearly all obsidian is heated for color, however, Mount Saint Helens stone fits right in with natural obsidian. In fact, through careful control of its manufacturing process, the simulant is often more appealing to the eye than its natural counterpart. Incidentally, because obsidian is glass, it obviously must be handled with care.

Rhodochrosite

For years, Afghanistan was the only source of rhodochrosite. It generally occurs in mass form as a translucent or opaque gem, though since the late 1800s beautiful crystals have been known to exist. Other sources include Australia, Germany, South Africa, parts of the former Yugoslavia, and Romania.

The United States is a major player too, thanks to recent developments in Colorado. The Sweet Home Mine, located near Alma, has given rise to cautious optimism in the trade. Although transparent material had existed there since late in the nineteenth century, the costs of extracting the material were prohibitive, keeping potential investors away, until now. This has vaulted the United States into position as the most significant source of rhodochrosite in the world.

Because rhodochrosite rates just 4.00 on the Mohs' hardness scale, it is seen more often in the creation of artifacts than jewelry. Its nearly perfect cleavage also makes it difficult to cut and facet. As a result, it is most often cut *en cabochon* and placed in settings that offer maximum protection for the stone. Rhodochrosite is more suitable for earrings, pendants, and brooches than bracelets or rings. Although it is hard to come by, rhodochrosite is nonetheless listed as an astral stone for the sign of Virgo.

When you talk to experts about the future of rhodochrosite, they often shrug their shoulders and freely admit they simply do not know. All known sources have proved to be unreliable, and rhodochrosite in crystal form is precarious at best. The Sweet Home Mine continues to yield some spectacular finds—for now—but making a projection is indeed like trying to predict the future in general.

Since consumer awareness is almost nonexistent, there is little demand for rhodochrosite at retail. As a result, it is almost never seen in conventional retail outlets. This fruitless cycle continues, keeping rhodochrosite from any place of prominence in the gem and mineral world. It is a most beautiful gem, though, and one I would strongly recommend if you happen upon it somewhere.

Rhodonite

Rhodonite is a pink-to-red gem that contains black veins of manganese oxide. This unusual color combination makes it quite easy to distinguish this stone from other gems and minerals. Because of its veins, rhodonite was once identified as pink turquoise. Obviously, this was eventually disproved, and rhodonite was reclassified.

Much of the world's supply of rhodonite is found on Vancouver Island, Canada, and in the Ural Mountains of Russia. Other significant sources include Australia, Brazil, Madagascar, Mexico, South Africa, and Sweden. In the United States, rhodonite is found in Massachusetts. At one time, rhodonite was the international gemstone of the Soviet Union.

It is most commonly found in an opaque form that is cut *en cabochon* or turned into lovely beads of pink and red. Once in a while, some rhodonite is found in crystal form, but that is quickly sold to savvy gem investors. It is never seen in a colorless state, for it always contains trace elements of other impurities.

Scapolite

Scapolite is most often seen in a yellow state, although it also can be found in shades of pink, gray, blue, and lavender, as well as colorless. Its sources include

Brazil, Canada, Kenya, and Myanmar. Because its structure is similar to that of the plagioclase feldspar, scapolite—particularly in its yellow state—is often confused with labradorite. To complicate matters, scapolite is also often found with deposits of feldspar. In its colorless state, scapolite has been confused with white sapphire and white topaz.

If conditions are right, scapolite can be seen cut *en cabochon* to display an asterism or cat's-eye effect. It also has been known to give off an eerie glow similar to that of moonstone. Scapolite is sometimes known as wernerite, after the noted and respected German gemologist A. G. Werner.

Serpentine

The mineral serpentine has been mentioned down through history as a gem-quality material found abundantly in just about every corner of the world. This is not true. In reality, the only gem-grade mineral in this group is a green-to-amber stone known as serpentinite. Because of its low specific gravity (2.60) and its poor hardness rating (2.50–5.00), some experts refuse to recognize even serpentinite as suitable for jewelry. Some of its more significant sources of supply include China, South Africa, parts of the former Soviet Union, Italy, and the United States. Serpentine rock also plays host to a wide assortment of other gems and minerals.

There are several varieties of serpentinite. The three most common are bowenite, a yellow-green variety that is sometimes mistaken for jade; williamsite, a dark green variety with black spots, which is sometimes found in the state of Pennsylvania; and verd antique, a dead ringer for dark green marble that often contains veins of calcite.

Serpentinite gets its name from the English word *serpent,* since it often resembles the skin of a snake. Because of this, certain earlier religions feared the greasy-looking material, believing its presence was a sign of the devil. Some cultures consider serpentine the birthstone for January. At one time, it was regarded as the international gemstone of Ireland, obviously in a connection between Saint Patrick and the snakes he drove out of the country. Today, however, Ireland claims the emerald as its official stone.

Serpentinite is found in beads or cut *en cabochon*. It can also be seen in artifacts carved by certain cultures from hundreds or even thousands of years ago to this very day. It is an interesting gem, but hardly worth a trip to Ireland to see it.

Sillimanite

Sillimanite is generally opaque in nature, though on rare occasions it may be found in a crystal form. For this reason, it is nearly always seen cut flat or *en cabochon*. Because it is found with fibrous inclusions, sillimanite is sometimes known in the jewelry world as fibrolite.

The most sought-after variety is blue sillimanite, found only in the ruby-rich Mogok region of Myanmar, while the gray-to-green variety comes from gem gravel in Sri Lanka. A third variety, found in massive rock form, is from the Clearwater River valley in Idaho; however, this material is not considered to be of gem quality, and it is normally used in the creation of artifacts. This same industrial-grade variety can also be found in Delaware. Consumers are unlikely to come across sillimanite at retail. In all my years as a gem salesman I have seen only two sillimanites, both of which were lifeless colors of green to gray.

Sodalite

Sodalite, together with lazurite, is a key component of lapis lazuli (see chapter 9). In fact, lazurite is often considered a member of the sodalite mineral group. Sodium oxide accounts for nearly 25 percent of the composition of sodalite; additionally, it has a low silica content and contains some chlorine as well.

Sodalite was first discovered in Greenland, where it was incorrectly classified as a variety of feldspar. It did not enjoy any popularity until a century later, when it surfaced in Ontario, Canada. This deposit ultimately became known as the Princess Mine after Princess Margaret, sister of Queen Elizabeth II of Great Britain. The princess was enamored of a beautiful blue stone given to her by her husband. Not surprisingly, sodalite is the international gemstone of Canada. Other significant sources of supply include Brazil and Namibia.

Titanite

Titanite (which is sometimes called sphene) may be seen in shades of green or yellow, either separately or in combination. In its yellow state, titanite is sometimes confused with heliodor (yellow beryl), yellow chrysoberyl, and imperial topaz. In its green state, titanite has often been confused with the demantoid garnet and chrome diopside.

Titanite is often found in rugged mountainous areas, within granite rock formations. California, New Hampshire, and Pennsylvania are three domestic sources of titanite. Other sources include Mexico, Canada, Austria, Switzerland, Madagascar, Myanmar, and Germany.

Crystals of titanite are quite small, and gems of more than two carats are virtually nonexistent. It is doubly refractive and strongly pleochroic, displaying at least three colors when looked at from different angles. Although titanite is normally seen only at gem and mineral shows, jewelers who specialize in custom and unusual work can sometimes create one as a special order for those who are in love with the stone. However, it is a brittle stone with a relatively low hardness rating of just 5.00, making it more suitable for rock hounds than jewelry lovers.

*W*eird Gems and Minerals

Before we move along, I did want to be certain to wring every last drop of information out of what's left of my mind. Why bother, you may ask. None of us will probably ever own—or even want to own—any of the minor stones left undescribed. Well, call me selfish if you want, but I didn't write this section of the book only for you; I also wrote it for me. Just in case I ever get picked to be on *Jeopardy,* I'm going to be ready! So here are some really weird minerals (really not generally up to gem grade), listed alphabetically; considering the contents, I thought it would be pointless to try to establish a pecking order.

Ammonite: A fossilized shell turned mineral, and actually a close relative of the squid. Why anyone would want one is beyond me. Found mostly in Morocco.

Artinite: Found in the gem-rich, fertile area of Staten Island, New York, it is defined in Webster's as a "hydrous magnesium carbonate in white orthorhombic crystals and fibrous aggregates." Go ahead: look it up!

Charoite: A mineral so complex that even scientists don't know how it is formed, except to say it somehow interacts with marble. Found only in a river that it's named after, in Yakutsk (part of Siberia), Russia. Take your long johns if you're going to get some.

Gaspéite: Olive green to brown, this material was originally found on the Gaspé Peninsula in Quebec, Canada. You should own one if you live there.

Herkimer Quartz: A variety of quartz named after the town of Herkimer, New York. One of the most pinpoint examples of gem location I have ever seen.

Lepidolite: A pink, bladelike silica that is actually a distant relative of the tourmaline. Found in Colorado and other parts of the United States, as well as Brazil.

Orthoceras: Another kind of weird fossilized shell found only in Morocco (see Ammonite, above).

Picasso: Guess the source of this name? Wrong! Picasso is actually some kind of marble that was named after somebody in Beaver County, Utah.

Septarian: A rather unsightly mass of yellow calcite and other minerals, this formed millions of years ago, when the Gulf of Mexico actually extended to the border of what is now Utah. I swear it's true.

Sphalerite: Semitransparent, this comes from the Chivela mines of Mexico and is sometimes known as zinc blende. I picked it, however, because it's also called "blackjack," which I think is a totally cool name.

Tinstone: Black material that is actually made of tin dioxide. Also known as cassiterite, it comes from somewhere in Spain, as well as Namibia. Tin prospectors value it highly. Why is anybody's guess.

After writing this section, I realized that the list of weird gems and minerals could literally fill a book unto itself, which sounds to me like a great idea—except that nobody would buy it.

Chapter 11

Jade

An Introduction to Jade

Jade is one of the world's oldest-known gems, dating back to the Stone Age. The earliest people depended on jade for weapons and other tools of survival. There are references in recorded history tracing jade to about 4000 B.C. It was easily the most significant stone of that time. In fact, jade came to be of such importance to the Chinese that some scholars have suggested there existed a Jade Age, between the Stone Age and the Bronze Age. Even today, certain cultures consider the stone sacred. And ever since ancient times, jade has been fashioned into objects of art and adornment that gem historians and archaeologists alike consider priceless.

Jade is regarded as a stone that imparts great wisdom, which makes it popular to this day with leaders of certain cultures. In fact, some Asian businesspeople keep a jade stone close at hand, believing it will allow them to make wise decisions that will result in great prosperity. Jade usually makes a person think of the Far East; however, certain groups in South America also consider jade to be sacred. In fact, cultures in both regions often bury their dead with a jade stone, to bring good fortune in the afterlife. Jade has been held to be a stone of much masculine energy, and for this reason some have thought it also could revitalize the dead. To this day, many people consider jade the "good-luck stone." Those who celebrate their birthday in May sometimes choose jade as their birthstone. Astrologers generally link jade to the planet Mercury, while individuals born under the sign of Pisces or Libra lay claim to it as their astral stone.

Besides the history and superstition connected to it, jade was also considered to have many medicinal qualities. Since jade is cool to the touch, some earlier cultures believed that it could control the fever of a sick child. Spanish conquistadors observed Mexican natives using jade, ground into powdered form, to treat kidney disease; for this reason, jade became known by the Spanish as *piedra de ijada,* or "loin stone." Jade was also believed to control ulcers, internal bleeding, and a wide variety of other internal ailments.

Myanmar is undoubtedly the world's most important source of jade. Others include China, Japan, South America, Mexico, and even the United States. Some interesting folklore is associated with prospecting for jade in Alaska and Wyoming. Fittingly, both also claim jade as their official state gemstone.

Shopping for Jade

There are two separate types of jade: jadeite and nephrite. They have much in common and equally share in the stone's folklore, history, and place of importance. They are distinctly different, however, in mineralogical composition, color, and appearance, so don't worry about having to distinguish one from the other. Jadeite is more translucent and usually a light apple green in color, while nephrite is a dark, almost forest green with little, if any, translucence. Each has its own appeal and significance.

The chief components of jadeite are aluminum and sodium, and its makeup (like nephrite's, but to an even greater degree) is said to be stronger than steel. In fact, one way to separate an impostor from the natural is to scratch the object with a piece of steel (such as a pen point or a pocketknife) in an inconspicuous place. If the steel object fails to scratch the stone, the chances are good that you are looking at genuine jadeite. The principal components of nephrite are calcium, magnesium, and iron.

Jade is measured in millimeters, rather than carat weight. If you are planning to spend major dollars on jade, it would be helpful to familiarize yourself with millimeters (there are just over twenty-five to an inch) so you can relate these numbers to the size of the pieces you are planning to buy. Check the stone for inclusions, most often black spots or veins that could potentially lower the value of your purchase when appraised. Since jade is nearly always polished, not faceted, you should be aware that too much heat generated by the polishing wheel can cause variations in color. Although this may sound negative in context, inconsistency in color can produce some interesting and unusual effects. After all, jade (for the most part) is not considered a costly gem, so don't worry too much about its value as a major factor.

The Importance of Color

Color is the single most important factor in determining the value of jade. Green is the dominant color in the group, but many others are available on the market, if you look hard enough. Let's try to help you narrow down your choices.

You may be surprised to learn that jade, when found in deposits without any impurities, is actually a fair and lovely white. Jade also comes in blue, yellow, red, black, and two distinct shades of green. Obviously green jade dominates the market, but this has more to do with availability and low consumer awareness than anything else. The fact is that many of the other colors are quite beautiful and well worth your time if you are really into jade. If you're lucky, conventional retailers may offer some green jade, but you'll have to search rock shops and gem shows to find any other color. Incidentally, jade is usually found in secondary deposits, and it weathers to a rather unattractive shade of gray if exposed to direct sunlight over long periods of time.

Jadeite

Most gem experts consider jadeite to be the only true form of gem-quality jade. Jadeite is finer in appearance and more translucent than nephrite. Jadeite can be found in a host of colors, including green, white, charcoal, red, golden yellow, brown, and others. The most highly prized jadeite of all is an emerald green gem known as imperial jade. This particular variety is much costlier than other colors of jade, and there are reports of dyed jade of lower quality and value that surface from time to time at retail. Before paying big bucks for imperial jade, it would be best to have it evaluated by a trained gemologist to be certain that the gem is natural, not dyed. If done well enough, a dyed piece can easily fool the untrained eye; however, natural imperial jade is colored by chromium, and it will display chromium lines when examined with scientific instruments. Myanmar is regarded as the world's only consistent source of gem-quality jade. Other, minor sources do exist, though, including China, Japan, Mexico, and South America, among others.

When shopping for jadeite in the retail world, you will most commonly find it cut flat or *en cabochon*. Since jade is translucent at best, little faceted jadeite is seen on the market today. Lovely bead necklaces and drop pendants of various shapes and sizes offer the jadeite shopper two other beautiful options. Believe it or not, there is a special variety of jadeite from mainland China that, when mined, is as clear as glass. Unfortunately, as soon as it is mined, the glasslike surface begins to deteriorate. This explains its lack of exposure and availability at retail.

Nephrite

Although it occasionally shows up in types of jewelry such as pendants, bead necklaces, and earrings, nephrite is primarily seen in objects of art, such as figurines, vases, and other decorative collectibles. It is sometimes also used in the carving of cameos, producing a most unusual, eye-pleasing effect. Like jadeite, nephrite is generally cut flat or *en cabochon*. Stunning, dark forest green beads also can be stumbled on in the jade market. Nephrite most often comes in a dark green color that is more vivid and opaque than its cousin, jadeite. Although its natural color is best described as eggshell white, nephrite appears in a number of other colors, including black, brown, and gold. It originates in many corners of the globe, including China, Japan, Mexico, Myanmar, and South America. In the United States, a deep green variety from Wyoming and a green-to-black stone from California have recently surfaced at retail. Since it is lower in cost than jadeite, these new creations can be found in a wide array of costume-jewelry pieces. Alaska and Wyoming have adopted jade as their official state stone.

Care and Cleaning of Jade

Jade can be cleaned in a number of ways. Obviously, start with the most conservative method—a soft polishing cloth—and go from there. A plain water mist would also be a safe bet, and a few drops of mild, detergent-free soap in a bowl of warm water should yield satisfactory results when all else fails. There is one essential point to make about the cleaning process for jade: do not soak the stone for more than a few minutes at a time. I would also stay away from harsh chemicals, steam cleaners, and ultrasonic units. After cleaning, an optional dab of olive oil will produce a wonderful shine.

A Few Final Words of Advice

Before leaving this section, I would like to touch on something I always have a problem with when I speak about jade: personal preference. Certainly I am not going to say that all jade, whether it be jadeite or nephrite, is the same, for top-quality, translucent jadeite is far more valuable than its darker, more opaque cousin. However, if you see a piece of nephrite jewelry that appeals to you, do not dismiss it just because the industry considers only jadeite to be gem-quality jade. Occasionally, I think we get so caught up in things like value and rank that they overtake our own preferences. That in itself is a crime. Personally, the color green is my all-time favorite, and as I write this section I am sitting in an office that has been painted a dark, nephrite green, with a trim that closely resembles the light green jadeite. I happen to love both colors, and I would buy a dark green strand of nephrite beads for my wife in a minute if I could find any at retail. Why? Because it appeals to me, and it makes me happy. Enough said.

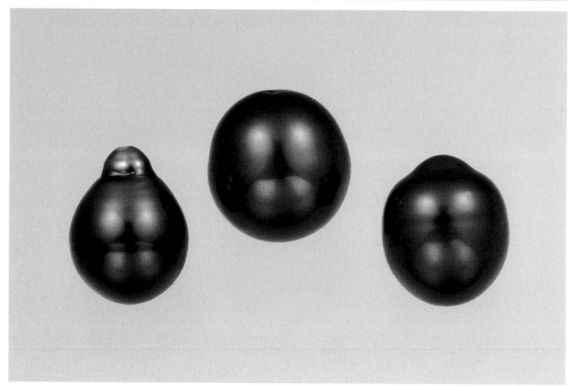

Chapter 12

Pearls

An Introduction to Pearls

No other form of jewelry has ever enjoyed such worldwide acceptance as the pearl. Whether natural, cultured, or created, a strand of pearls has always been viewed as the one essential part of the jewelry wardrobe that will never lose its appeal or fall victim to changes in style. In fact, it is this ability of the pearl to adapt to virtually any new fashion that has fueled its popularity down through the ages. In terms of sales, pearls rate second only to the ever popular diamond.

The pearl is the oldest-known gem, and for many centuries it was considered the most valuable of all. Unlike virtually all gems, the pearl is actually organic matter, derived from a living creature—in this case, oysters and other underwater animals called mollusks. Much has been written about the folklore and tradition of the pearl. In fact, so much history accompanies the pearl that it has been claimed as the birthstone for no less than five months: February, April, June (traditional), July, and November. The pearl is also an astral stone for the signs of Gemini and Cancer, and astrologers link it to the moon. Couples celebrate their thirtieth anniversary with the pearl. Some early cultures believed a single drop of rain once fell from the heavens and became the heart of the oyster, thus bringing forth the birth of the pearl. Others referred to pearls as the "teardrops of the moon," speculating that oysters were drawn to the surface of the water by the rays of the moon and fertilized by droplets of dew when they opened their shells. Still others popularized the notion that pearls were formed by the passage of angels through the clouds of heaven. Early civilizations spoke of the medicinal properties of the pearl, employing it to treat everything from indigestion to heart disease. Actually, in some instances, they were not all that far off: calcium carbonate, its primary substance, is used even today as an antacid.

Over time, the pearl has become the symbol of purity and innocence, and as such it is considered an ideal wedding gift. Many bridal gowns incorporate it into the bodice of the dress, and a strand of pearls, together with similar earrings, has become an almost essential part of the bridal wardrobe. In certain religions, an undrilled whole pearl is presented to the groom during the wedding ceremony as a symbol of chastity and grace.

Pearls can be found in a variety of sizes and shapes, and they may occur either naturally or through a process known as cultivation. Although the oyster is by far the most widely used mollusk in cultivation, the pearl can be found within a wide variety of other species, including the mussel, clam, abalone, and conch. Natural pearls are very seldom seen and, as you might suspect, prohibitively expensive. The cultured pearl has taken over to fill the gap. Cultured pearls are often separated into two subgroups: saltwater and freshwater pearls. During this discussion, pearls that are mentioned without any designation should be considered saltwater or sea pearls; freshwater pearls will be identified as such.

Natural Pearls

Natural pearls are those that are formed without any human assistance. Basically, a natural pearl is created when a grain of sand, piece of shell, or other tiny bit of matter invades an oyster, clam, or other mollusk. As a protective device, the oyster emits first a black coating known as conchiolin (pronounced "cong-kye-eh-lin") and then a white nacre (pronounced "nay-ker"), which surround the intruder and thereby eliminate the source of the irritation. The black conchiolin helps the nacre adhere to the irritant. These oysters are gathered from great depths by divers who are sometimes known as pearl hunters, or by large nets called dragnets. Out of thousands of oysters that may be gathered at one time, only a minute fraction contain pearls at all, and an even smaller fraction yield pearls that are worthy of gem-quality status. Believe it or not, the ratio of oysters actually bearing natural pearls to oysters harvested may be as low as one in ten thousand! Because of the rarity of the natural pearl, a single strand of them can command prices in the tens of thousands of dollars or even more in the retail marketplace, putting these gems well beyond the reach of the average consumer.

The pearl oyster resides in various oceans and seas, most notably the Persian Gulf. The world's oldest still-active pearl fisheries are those in the Indian Ocean that grace the coast of the gem-rich island of Sri Lanka. These fisheries have been providing beautiful-quality pearls to the marketplace for more than two thousand years! Some exquisite specimens of natural pearls also can be found within the shells of the abalone in the blue waters of the Pacific. These colorful specimens are irregular in shape and can be multicolored. They often display a dazzling effect known as pearl essence and sometimes almost seem to glow. Unfortunately, because of a shortage of abalone, commercial harvesting at this time is forbidden. A bill designed to save the red abalone mollusk that produces the abalone pearl has been introduced in Congress. This measure, if passed, would ban the harvesting of red abalone south of San Francisco, closing the fisheries and leaving this area to sport divers. In general, the natural pearl has been pushed to the brink of extinction, as a result of overfishing earlier in the twentieth century. Adverse climatic conditions and water pollution connected to the industrialization of Far Eastern nations have both complicated this situation. I suppose with progress comes a price. These problems are not expected to change anytime soon; in fact, some predict that conditions will worsen as we enter the twenty-first century.

Cultured Pearls

The first cultured pearls grown in modern times were attributed to the legendary Kokichi Mikimoto of Japan in 1893. He applied for and received a patent on the process some twenty years later, but the market was hardly waiting with open arms. Major jewelry dealers and retailers refused to classify the pearls as real, and they even established an "anti–cultured pearl" policy to unify the opposition. Needless to say, this opposition didn't last, and today consumers worldwide are proud to display their cultured treasures. Unquestionably, cultured pearls now dominate the pearl market.

The Cultivation Process

An oyster creates a cultured pearl in much the same manner as a natural pearl, but under a controlled environment induced by humans. In this process, called nucleation, a small piece of live tissue is taken from an oyster and placed around a tiny bead that is harvested from a mollusk known as a pig-toe clam. The irritant is subsequently inserted into another oyster, which is then gently lowered close to shore for a period of not less than two weeks, so that it can adjust to its new life and purpose.

The mother oyster generally takes from one to three years to yield a mature pearl. Actually, a saltwater oyster can bear as many as five pearls at a time, but the oyster has little chance of accomplishing that prodigious feat. Because the oyster is such a delicate organic, cultured-pearl farmers observe and protect each one, closely monitoring water temperature and keeping away potential invaders. As you can imagine, this is a continuous process. The oyster reacts the same way it does when forming a natural pearl, surrounding the bead first with black conchiolin and then with thin, pearly white nacre. (Although some believe the pearl is almost entirely nacre, the fact is that the white substance forms an extremely thin layer. Science has proved that, in general, the nacre accumulates at a thickness of only .15 millimeters per year, meaning that a two-year-old pearl has just .30 millimeters of nacre, a three-year-old has .45 millimeters, and so on. As noted in the previous chapter, it takes a little more than 25 millimeters to make up a thickness of one inch.)

Extreme variations in temperature, longer-term climatic conditions, tidal waves, pollution, and predators all combine to destroy untold numbers of mollusks that could potentially bear a pearl. As much as 50 percent of the fragile akoya oysters die soon after implant, for example, and of those that survive, only one in five will ever actually yield a pearl. Keep in mind that these harvested pearls then may or may not be judged suitable for production. As you can imagine, all these potential problems subject pearls to price fluctuations.

Akoya Pearls

The akoya oyster is responsible for about 80 percent of the retail sales of pearls worldwide. First discovered and cultivated by the Japanese in the nineteenth century, this small, delicate creature takes three years to mature. Akoya oysters generally produce pearls that range from two to ten millimeters in diameter. Although their optimum shape is round, they are found in just about every other kind you can imagine. Since perfectly round pearls are seldom seen, however, the shape most sought after is known in the industry as "near round." The pearls are mostly white, but at times the akoya oyster can produce pearls of cream, gray, blue, green, or pink. The akoya oyster is held in highest regard among pearl cultivators, because it can produce pearls that are often the finest quality in the world.

The cultivation of an akoya pearl is done in the exact manner described above. The longer it is left suspended in the water, the thicker the nacre becomes. Unfortunately, greedy pearl farmers sometimes pull pearls up way before their time to meet the ever-increasing market demand. Pearls that are harvested before their time yield a much thinner nacre, with considerably less luster (or inner glow). Still, this does serve a purpose: to allow consumers who normally would have to settle for imitation pearls to own cultivated ones instead.

Cold-water pearls like the akoya yield a higher luster than those from the South Pacific. This is because cold water slows the rate of secretion, which in turn increases the luster. Harvesting takes place during the colder months of the year, generally from October through March if a good crop is successfully cultivated.

Although the Japanese continue to dominate the market, increased production in China has loosened their hold on the industry. Production is up across the board—which is good, since consumption in the United States is projected skyward for the future. Thanks to increased demand, higher prices had already been predicted when a new factor dramatically affecting supplies surfaced in 1997. In that year, the global pearl harvest was approximately 25 percent lower than in the previous year, as a result of a mysterious virus. Add that to the high demand for pearls worldwide, and as you might expect, price increases of 15 to 20 percent are now commonplace in the industry. Even though many merchants still had goods purchased from the previous harvest at lower costs, they were more than happy to raise their prices quickly on older goods as well.

Mabe Pearls

Another variety of cultured pearl currently enjoying great popularity at retail is the large mabe (or half) pearl. Mabe pearls (sometimes referred to as blister pearls) are made by affixing a bead that can be composed of plastic, soapstone, or mother-of-pearl (the substance lining the inside of a mollusk's shell) to the lip of a large oyster, which then secretes nacre around the irritant. The hemispheric pearl is eventually cut or stamped out of the shell and thoroughly cleaned to prevent deterioration. Next it is filled with a resin and closed with a mother-of-pearl back. Most mabe pearls are cultivated in the species called *Pinctada maxima*, the same giant that is found in the South Pacific. I have personally held one of these magnificent creatures in my hands, and I can tell you this: it's not exactly something you would order as an appetizer at your favorite fish house. These oysters may contain five or more beads per shell. Another variety that contributes to the mabe-pearl supply is known as the penguin-wing oyster. Cultivated in southern Japan, this oyster yields a more colorful and sought-after variety of mabe that is encountered less frequently and costs more. Although the round mabe pearl is the most popular and most common, it can be produced in a wide variety of other shapes—among them hearts, teardrops, and ovals—merely by altering the shape of the nucleus.

Cultivation of mabe pearls requires considerable attention and labor; nevertheless, they continue to be affordable for most consumers. Large, lovely mabe pearls of twelve to fifteen millimeters can generally be found in 14-karat-gold mountings for less than $200. The mabe pearl is also often used as an enhancer for the cultured-pearl necklace.

Mabe Shell

Before leaving this section of the book, I believe you should know that there is another cultured organic making headway in the market. Generally known as mabe shell, this product is identical to the mabe pearl in every way but one: it does not contain the nucleus of a pearl or bead. It is, in fact, simply mother-of-pearl. Essentially, the mabe shell is a by-product of the cultivation process. When the oyster secretes the mother-of-pearl used as a protective measure, it does so not only around the pearl but also on the shell itself. This mother-of-pearl, which clings to the shell, is later extracted and fashioned into a mabe shell. Removal of the material is done either by stamping it out using a mechanized punch or by carving it out of the shell by hand, obviously a more tedious and costly method. On the surface, the mabe shell is a dead ringer for the mabe pearl; the only way to tell them apart is by examining them with an X ray. Since the mabe shell is generally about half the price of a mabe pearl, or even less, it makes an excellent option for the more budget-conscious consumer who wants the look and feel of the same cultured-pearl product.

Other Varieties of Pearls

The conch pearl made its debut in the consumer marketplace in the summer of 1996. Formed naturally inside the mollusk of that name in the warm waters of the Caribbean, this new beauty can be found in colors ranging from a lovely shade of pink to a golden brown variety. Because its natural coating has a different makeup from that of the natural pearl, the conch pearl cannot technically be classified as a pearl. Sometimes confused with coral, the conch pearl is usually quite small in size. Most experts agree that there is little chance the conch pearl will ever approach the popularity of the other varieties of pearls.

Another variety you may encounter from time to time is the keshi pearl. Although it results from a natural intrusion, it forms while the oyster is in the cultivation process. The keshi pearl is a by-product of that process, and so it is classified as a cultured pearl. Such pearls are usually quite small and irregular in shape; still, they offer the pearl fancier another affordable option. Keshi pearls may form in any variety of mollusk but are most valuable when found in the large South Pacific oysters. These pearls can also be referred to as poppy or seed pearls.

Pearls of the South Seas

South Sea pearls are the most valuable of all pearls today. They are more popular than ever before, triggering price increases across the board. Consumers are not fazed by this, however, and worldwide demand just seems to keep growing.

The fertile waters off Tahiti yield many highly prized pearls, but they can be harvested in other areas of the South Pacific, as long as the waters are clear and pure, with a mean temperature of about seventy-five degrees Fahrenheit. These pearls are also found in the waters off Australia, Myanmar, and the Philippines. South Sea pearls grow in a mollusk that is the largest of its kind in the world: the *Pinctada maxima,* or gold lip. These giant oysters can measure as much as a foot across or more and weigh anywhere from ten to fifteen pounds! As a result, the warm seas yield pearls that can be up to fifteen millimeters in size, compared to the ten-millimeter maximum of cold-water pearls like the akoya.

Besides their size, South Sea pearls are valued for their exquisite array of colors. Most sought after is the golden pearl, which grows within the gold lip in the more northern seas off Southeast Asia. Pearls with overtones of silver also are highly desirable. As you might guess, this variety grows in the closely related oyster called the silver lip, found mainly in the waters off Australia. A black pearl comes from another *Pinctada* species, called the black lip. Still other available colors of South Sea pearls include cream, pink, green, and blue. They are truly treasures for the eye to behold.

The South Sea pearl has been cultivated since the 1950s and is easily distinguishable from the akoya pearl. The two also are oceans apart when it comes to price. While akoya pearls are affordable for the average pearl shopper, jewelry made from the South Sea pearl fetches thousands of dollars or more at retail. Its trendiness continues, and forecasters were predicting a growth rate that could reach 30 percent in 1998. Although the demand for the akoya pearl is also growing, it cannot keep pace with the market for the South Sea pearl.

Cultured Black Pearls

One of the hottest trends in jewelry today is the natural black pearl. As we have just seen, consumption rates for all pearls are on the way up. Needless to say, so are prices. The black-pearl market is dominated by Tahiti, though the waters off the coast of Okinawa enjoy a small share of supply too. Although they are referred to as black pearls, these beautiful specimens may actually display undertones of blue, green, and violet, or a mixture of colors in a spectacular creation known as the peacock pearl.

Natural black pearls lead to some of the most expensive pieces of jewelry on Earth, with large necklaces commanding hundreds of thousands of dollars or more. These pearls are so rare that it often takes a dealer ten years to assemble enough matching pearls to make even one single necklace! Admittedly, most of us will probably never own an entire strand of natural black pearls, since they are well beyond our reach. Still, there are alternatives. Single, large cultured black pearls have recently appeared at the retail level, often set with diamond accents as beautiful pendants and earrings. Generally speaking, such pendants can be found for $300 and up, with rings and earrings considerably higher in price. Obviously, the cost is relative to the luster, color, shape, and surface feel. Necklaces, however, are pretty much the property of wealthy investors, usually purchased through high-end retailers or a loosely organized network of dealers, importers, and exporters.

As if such pearls weren't rare enough, the forces of Mother Nature have put a strain on this already-volatile market. A massive tidal wave more than seventeen feet high slammed into the southwestern islands of French Polynesia in July 1996, causing massive damage to many South Sea cultured-pearl farms along the way. Many of Tahiti's farms of fragile black pearls were damaged or destroyed on impact by the wall of water. So severe was the devastation that experts now predict smaller farmers with centuries of family-owned tradition may be forced out of the business for life.

Add this unfortunate tragedy of nature to the cultured black pearl's current increase in popularity, and the result is that the price of this pearl is heading in only one direction: up. If you are planning to add even a single cultured black pearl to your collection, you should do so at the first opportunity, for the demand and the asking price have shown no signs of weakening.

Freshwater Pearls

Natural Freshwater Pearls

Natural freshwater pearls and natural saltwater pearls are formed in much the same way, only in different bodies of water and within different varieties of mollusks. Fine-quality natural freshwater pearls of good color can bring hefty price tags, regardless of location. Beautiful specimens of fifteen millimeters or more can fetch thousands of dollars and up at retail.

At one time, the fresh waters of North America were a major player in this market. Records dating back hundreds of years indicate that Native Americans wore them as amulets and necklaces long before the arrival of Europeans. Early Native American traders knew little of the value that these newcomers placed on their finds, and natural pearls worth hundreds or even thousands of dollars were sold for practically nothing. In addition, these earliest pearl prospectors completely disregarded the value of the shell, considering it a byproduct of their labors. Today these shells are sold for use in the manufacture of decorative buttons for clothing.

Cultured Freshwater Pearls

The start of the cultured-freshwater-pearl industry dates back about six hundred years or more, when the Chinese first discovered that a small pearl could be cultivated within a large mussel in lakes and rivers. Initially, crude irritants such as wood and metal were used to induce the mussel to create the pearl, but this practice eventually disappeared as cultivators began to utilize a section of shell or a piece of another mussel. Although the process was deemed a success, pearl farmers later found that by incorporating a round mother-of-pearl bead instead, they could induce the mollusk to bear a freshwater pearl that was rounder and more symmetrical in shape. Over time, cultivators have found that varying the shape of the mother-of-pearl implant could produce other shapes such as the oval, the bar, and the marquise.

When the average American thinks of a mussel, a picture of a small juicy specimen inside a black-to-purple shell comes to mind; actually, the mussel used to cultivate the freshwater pearl can grow up to a foot in length and more than eight inches across! Because of their massive size, these mussels can produce as many as thirty pearls in one shell. Normally a consumer will encounter sizes ranging from 1.5 to 6 millimeters, but larger specimens occasionally may also be found at retail.

Cultured freshwater pearls are most commonly found in varying shades from white to cream. Avid freshwater-pearl buffs can locate their treasures in other colors as well, including pink, orange, violet, gray, and blue. Like the saltwater pearl, the freshwater pearl is seldom seen in a perfectly symmetrical, round shape. For this reason, the classic necklace of cultured freshwater pearls incorporates the near-round shape.

For many years, Japan was the worldwide leader in the production of cultured freshwater pearls. Known in the trade as Biwa pearls, these specimens are cultivated in Lake Biwa, the nation's chief pearl-producing source. Times change, however, and the Biwa pearl is not nearly as dominant as it once was. These days, China is generally recognized as the industry leader, but the United States is closing fast. In particular, production in Tennessee and Arkansas has exhibited substantial growth, and overall the domestic freshwater-pearl industry is expanding. Still, American market share is quite small when compared to the production coming out of China. Nevertheless, the increase in domestic production is eroding the once-healthy profit margin of the Chinese goods.

The outlook for the cultured freshwater pearl is encouraging. Prices and supplies have stabilized as the Chinese increase their carefully controlled production to keep up with increasing consumer demand. Smaller pearls can still be found at very attractive prices, and all shapes except the larger rounds are expected to remain in good supply.

Simulated Pearls

A staple of the costume-jewelry industry, simulated pearls are made in a variety of ways. They can be made of plastic, glass, soapstone, or other products, and, as you can imagine, prices reflecting various levels of quality are all over the place. Some of the highest-quality and most appealing imitation pearls are those coated with a product known in the manufacturing world as pearl essence. This product is a liquid used to coat an imitation pearl made of glass or plastic. This application adds a lifelike top-quality luster that is less distinguishable to the naked eye than lower-end competitors. Interestingly, pearl essence has absolutely nothing in common with a pearl. Most people are surprised to learn that this coating is made not from a pearl, mollusk, or shell, but rather is derived from the scales of a certain type of herring. (It's not surprising that this product does *not* go under the name "herring essence"!) Still, these imitation pearls, when manufactured with care, can offer an affordable alternative to the cultured pearl. Some of the best simulated pearls come from the Spanish island of Majorca.

Since the pearl is such a delicate creature, it would be best for you to add a strand of top-quality simulated pearls to your wardrobe if you are planning on wearing them often. Obviously, simulated pearls are much more durable than their cultured counterparts, and they make an excellent alternative for daily wear. When pricing simulated pearls, do not hesitate to ask the salesperson the composition of the pearls you are considering.

Shopping for Pearls

Shopping for pearls can prove to be a formidable task for even the most ardent jewelry collector. There are so many choices and arbitrary colors that consumers often take months to make their decision. This in itself is not a bad thing, but let's see if I can pass along some hints that help move things along.

As we have already seen, the quality of a pearl is as widespread as the quality of a diamond or any other gemstone. Factors that affect the value and quality of a pearl, in order of importance, are luster, size, shape, surface, and color. Carefully consider all of these factors, for each is extremely important.

Luster

Although no one factor dictates the value of a pearl, most experts agree that luster is at the top of the list. Luster can be loosely defined as the reflection of light off the surface of the pearl. This reflection is important, but pearls of excellent luster also seem to glow from within. Some people even think this glow actually changes to reflect the mood of the wearer. Whether it be bright and white or warm and cream, Mother Nature eventually determines the luster of a pearl. Pearls showing pink overtones generally have the most desirable luster of all. Consumers should also keep in mind that pearls are sometimes irradiated to enhance their luster.

Size

Pearl sizes are measured in millimeters, and their weight is expressed in an extremely small unit of measurement called grains. As a point of reference, keep in mind that a five-millimeter pearl will weigh approximately 3.5 grains. As you might expect, larger-size pearls are worth far more than smaller ones of equal quality. Generally speaking, pearls that retail for $400 or less tend to be five to six millimeters in diameter. As the pearls get larger, the cost rises accordingly. If you want the appearance of large pearls but cannot part with thousands of dollars, a graduated pearl necklace (in which the size of the pearls increases around both halves of the strand) would be a lovely choice. If you decide to purchase one, remember that the pearls should have a slow transition in size, and opposing pearls should match.

As stated earlier, about 80 percent of all worldwide retail sales of pearls come from the akoya oyster, so there is a pretty good chance that the cultured pearls you own or are planning to purchase are akoya pearls. As we have also seen, the akoya pearl seldom, if ever, exceeds ten millimeters in diameter, while at times South Sea pearls can be as much as fifteen millimeters across.

Shape

Pearl shapes are classified as either classical or baroque. Classical pearls, or perfectly round ones (as well as fancy rounded shapes such as the teardrop and the oval), are the scarcest and most valuable of all. The round (or, as we have learned, near round) is also by far the most popular shape. When shopping for round pearls, make certain all the pearls in the piece under consideration are approximately the same diameter and shape.

Pearls found in irregular shapes are generally classified as baroque. It is important to note that this terminology refers only to those pearls with a distinct shape that is clearly other than round. Pearls that are slightly off from being perfectly round are generally not considered baroque. Pearls with unusual names and shapes to match can also be found on occasion at retail. For example, potato pearls really do resemble the potato, corn pearls look as if they just came off the cob, and rice pearls seem good enough to eat with your Sunday dinner. Although these shapes provide a look all their own, to me none of these shapes can ever replace the simplicity and the beauty of the classical, near-round pearl.

Surface

It is said that perhaps one pearl in a million makes the grade as flawless. Remember, pearls are produced by nature, and each has its own little fingerprint. In fact, this fingerprint is your best assurance that your pearls are genuine, not synthetic or simulated. In some instances, however, I have observed costume pearls with sand incorporated in the manufacturing to deliberately flaw them and add credibility and a more realistic surface than those that are as smooth as glass (or plastic). These tiny specks of nature are a vital part of their composition and do not detract from their beauty and value unless the damage is very obvious and excessive.

When shopping for pearls, be sure to closely examine the finish or surface of the pearl. Check each one for rough spots, cracks, or any other defects. Rub each pearl gently between your index finger and thumb, looking to see if the finish of the pearls is smooth and soft to the touch.

Color

Although pearls may be found in a wide variety of colors, the most prized ones are white, cream, rose, and black. Rarely you may encounter other shades, such as yellow, blue, green, brown, and gray. For these colors to have an impact on value, keep in mind that they should be natural, not dyed. Consumers should also know that pearls are sometimes bleached if the black conchiolin is visible through the nacre of the pearl. In addition, if the pearl has a striped effect, it is showing the growth pattern of the bead. This should tell you that the nacre is too thin. It should also tell you to look for another strand of pearls. When making your purchase, request some documentation from the salesperson to confirm that your cultured pearls are not enhanced in any way. Although pearls are available in a whole rainbow of colors, they all have at least one thing in common: consistency. Whether you choose white, rose, green, black, or any other color, be certain that the pearls are all similar in color, for poorly matched strands can result in a sharp reduction in value.

Getting Your Money's Worth

Shopping for pearls can be a tricky and sometimes bewildering process, and pearl quality may prove to be more difficult for the average consumer to determine than that of colored gems or even diamonds. Some simulated pearls are of such high quality that today's average pearl buyer may find them an additional source of confusion. Although you will probably never be 100 percent certain, there are some things you can do to protect yourself and get your money's worth.

For starters, always examine pearls against a background of solid color, rather than on top of a glass display case. The solid background will better enable you to see the luster, the number one factor in determining the value of a pearl. Take your time, and study each and every pearl individually. If the retailer has more than one strand on display, don't hesitate to check one against another for quality, because pearls may exhibit even subtle differences that could affect their market value. Don't forget to evaluate the clasp too, for practicality and beauty. Although most pearl necklaces and bracelets of any value incorporate a 14-karat-gold clasp, it would be a good idea to make certain that the piece is stamped and easy to use. In order to prevent delicate pearls from rubbing against each other, most manufacturers will use a double knot on a silk cord between each. Silk is the material of choice, for its smooth texture. The double knot should be easily apparent to even the least-experienced consumer. Needless to say, this double knot is also an important extra security measure, as well as a sign of quality.

Since natural pearls are rare and extremely expensive, it is a safe assumption that the pearls you are evaluating are cultured, even if the salesperson tells you they are genuine. Unfortunately, not all cultured pearls are of the same quality and price, and the differences may not be as apparent as you might expect. While keeping in mind the factors of luster, size, shape, surface, and color will certainly help, your best assurance of all may be the retailer himself. Shop with established, reputable sources and your chances of being taken in will be minimal at best. Most often, searching for that bargain while vacationing in unfamiliar territory could prove to be your tragic flaw.

Since pearls are customarily strung on silk, it is likely that they will need to be restrung somewhere down the line. In fact, some pearl experts recommend you get them restrung every year. Many retailers offer this service, though few, if any, will do it free of charge beyond the warranty period. Still, as a point of information, it's a good idea to see if your retailer offers this service. Be sure the dealer clearly explains the store's return policy, and get the store's guarantee in writing. If you paid a considerable amount of money for your pearls, do not hesitate to have them appraised by a certified gemologist. If you've carefully done your research and have been certain to pay attention to detail, shopping for pearls should prove to be a most pleasant experience. In closing, remember my earlier warning: as with most other things in life, if the deal seems just too good to be true, chances are it probably is.

Disclosure: Policing the Industry

The current standard for determining quality in the pearl industry was developed by the Japanese government in 1952. A group established minimum quality-control standards for all pearls that Japan exported worldwide. I suppose in its time it was hailed as revolutionary, but times have changed, consumers have changed, and it's time for the industry to do the same.

As of January 1, 1999, the industry was scheduled to implement a new, updated inspection system. This system upgrades industry criteria and standards, examining five major areas in particular: nacre thickness, luster and clarity, surface imperfections, nacre damage, and processing methods used, if any.

One major drawback is that exporters do not have to participate; it is entirely voluntary. So how effective can such a system be? Well, the Japanese Pearl Exporters Association (JPEA), charged with implementing the system, has some very intense marketing plans for the program. The JPEA made plans to launch a worldwide marketing campaign, beginning in early 1999, aimed not at the dealer but at the consumer. The hope is that by making consumers aware of its existence, the JPEA will influence them to demand that the pearls they are considering pass these quality-control tests. All pearls that the JPEA examines will have a tag identifying them as such. It is hoped that better-educated consumers will demand only pearls that have made the grade, and that this in turn will gradually force exporters, importers, and retailers alike to carry only the tagged pearls. The JPEA should be applauded for taking this aggressive stance, and I for one hope consumers back it wholeheartedly. If demand for the tagged pearls is there, the system will work.

Care and Cleaning of Pearls

Always remember that pearls are fragile, delicate organics, and therefore must be treated with extreme care. As a rule of thumb, your pearls should be the last thing you put on, unless it's February and you live in North Dakota, in which case your overstuffed 100 percent goose-down ski parka should be the last thing you put on. Apply all makeup, perfume, and hair spray before putting on your pearls. Be certain not even to lay them in the vicinity when you spray or spritz, as certain harsh chemicals and perfumes can damage the surface of your pearls or diminish their luster forever. Keep your pearls stored in a separate compartment or soft pouch, away from any contact with chains, rings, or other jewelry items that can easily scratch their surface. While paying so much attention to the pearls themselves, don't forget to take care of the silk cord they are strung on.

Silk is delicate too, and it will eventually stretch or give way if exposed to chemicals and other liquids. It is good advice to check your pearls every time you wear them to be sure the silk cord shows no unusual signs of wear. If you follow this practice, you may avoid the heartache of losing one—or all—of your pearls while out for the evening.

Although I have seen some liquid cleaners in the marketplace that claim to be safe for pearls, personally I would take no chances. A soft cloth made of silk or a similar fabric is your best choice when cleaning pearls. If you notice a foreign substance such as makeup during the cleaning, your best option is to use a gentle solution of warm—not hot—water and a mild nondetergent soap. When drying your pearls, be certain to lay them flat on a piece of soft material that is free of dyes and detergents. You should also store your pearls in this same manner. Do not hang your pearls for either drying or storage; the pressure put on them could cause them to break apart or stretch.

If you notice a more stubborn stain that conventional cleaning methods do not remove, consult your local jeweler or gemologist for advice. Unfortunately, because pearls are porous, they can sometimes absorb a substance that is impossible for even an expert to remove, and replacing the pearl (or pearls) involved may prove to be your only viable option. If this should happen to you, be sure the total cost involved is clearly explained and given to you in writing. Chances are that your pearls may have to be shipped elsewhere for repairs, and this could possibly alter your decision as well. Never leave your pearls behind, for this or any other reason, without a written receipt from the dealer. I also recommend that whenever you leave your pearls behind for replacement, restringing, or any other reason, count the pearls on your strand and record that number before handing them off to your dealer. You could even ask the dealer to do the same. After all, it is a way of protecting both of you.

Dan's Dissertation on Pearls

The pearl offers a look and a feel that no other gem, not even the diamond, can duplicate. Pearls are feminine, soft, pure, and understated. Even if you are on a limited budget, an eighteen-inch strand of cultured pearls of acceptable quality can still be found for less than $300 if you scour the market for sales. The special feeling that pearls create is well worth the price, even if you need to save your pennies to own a strand. In my opinion, every woman should own a strand of cultured pearls. If you're single and you can't afford much, sell something useless like your washing machine to buy one. If you're married (or otherwise involved) you're really in luck, because you can raise the money by selling something even more useless, like his bowling ball or his golf clubs.

Chapter 13

*O*ther Organic Groups

*B*ackground Information

The pearl is unquestionably the most popular organic gem; in fact, it dominates the organic group just as the diamond dominates the gemstone market today. Yet the pearl is not the only member of this group that is suitable for jewelry and available on the market today. The others are amber, coral, and jet. Ivory and tortoiseshell are also organics, but both are protected by law and seldom seen in the jewelry trade anymore. They have been replaced by simulants composed of fossilized bone, plastic, resin, and a wide assortment of other materials.

Although organic matter technically cannot be included in the category of gemstones, in no way does that diminish its importance in the jewelry world. In fact, organics have been around since the beginning of life on Earth. Since the pearl has already been discussed in detail, our focus in this chapter is on the remaining three significant organics, followed by a brief discussion on ivory and tortoiseshell.

*A*mber

Amber is technically not a gemstone or mineral, but instead is fossilized sap from prehistoric trees that grew as far back as almost fifty million years ago, primarily in Scandinavia and elsewhere around the Baltic Sea. Today the most important source of this most unusual organic substance continues to be Russia, in particular the Kaliningrad mine. The area of Poland along the Baltic coast, not far from Kaliningrad, is another major source; in fact, occasionally amber can be found floating on the surface of the Baltic, after being set free from land by the constant pounding of the sea. The average consumer thinks of amber only in its golden yellow state, but it also comes in shades of milky white, red-orange, green, black, and even (rarely) violet. It is generally accepted that the amber from the Baltic region is the world's finest. Actually, some gem purists consider it the only true amber in the world today. It is usually seen cut *en cabochon* and polished to a high shine. Amber is also sometimes carved into beautiful and unusual pendants, brooches, and other such pieces of jewelry, as well as decorative artifacts. (Possibly the most highly prized of all the art treasures that were looted during World War II and are still missing is the famous Amber Room in the palace of Catherine the Great in Pushkin, outside Saint Petersburg, Russia. Not only the luxurious paneling but even all of the furniture and decorations were made entirely of carved amber.) An artist will carve an image into the flat side of the cabochon, and the transparent nature of the amber will magnify it and allow it to show through.

Amber ranks at the bottom of the specific-gravity scale, being measured at only 1.08. Because of this, huge pieces represent tiny carat weights, so amber is normally measured in millimeters instead. The heaviest piece of amber on record weighed just twenty pounds. Amber is popular in other ways besides jewelry: an oil derived from it is used in the production of certain pharmaceuticals, and because it gives off the aroma of musty pine when heated, certain cultures use it in incense. If the opportunity ever presents itself, visit the Museum of Amber in Sopot, Poland, a thriving community on the Baltic Sea. If you do, you're sure to smell the amber burning in strategic places throughout the museum.

With its usual golden yellow color, amber is a nontraditional birthstone for November. Some people born under the sign of Leo choose amber as their astral stone, but there are also references to it as an astral stone for the signs of Aquarius, Pisces, Cancer, and Scorpio. Amber is also considered the tenth-anniversary commemorative stone.

Amber probably has more folklore and tradition associated with it than any other gem. It has been spoken of with reverence by ancient tribal leaders and medicine men for its claimed healing properties. Early physicians prescribed amber as a cure for headaches, arthritis, heart problems, and a variety of other ailments. When embarking on a long journey, many travelers wore amber amulets to protect them from fatigue, heatstroke, and death. Christians thought that finding amber indicated the presence of God. In the Far East, amber is regarded as a symbol of courage; in fact, certain Asian cultures still believe that amber is actually the soul of the tiger. Egyptian cultures tucked pieces of amber into the caskets of loved ones, in the belief that the body would then forever remain whole, even as it was transported to another life.

From time to time, tiny bits of ancient plant and animal life have been found trapped inside amber, leading some people to consider it a key to information about the beginning of the world, a type of organic matter that transcends time. Scientists have even managed in recent years to draw and examine DNA from insects preserved in amber. Perhaps most astonishingly, a piece of amber containing a fully preserved flower complete with seeds was discovered in the Dominican Republic in 1995. This find is believed to be the first of its kind, and it could eventually lead to the reconstruction of a tree that has been extinct for more than fifteen million years! On the other hand—except in Hollywood—modern technology is still unable to create a dinosaur from the animal blood that has been found inside some of those trapped insects. All this aside, amber is a most mysterious, mystical organic gem that should be a part of your jewelry wardrobe. It is not particularly expensive, and it sometimes can be found set in sterling silver, either alone or with marcasite. Amber is readily available, and there is no expectation of a shortage.

Coral

The result of an accumulation of skeletal masses, coral has been called by some "the garden of the sea." Until the eighteenth century, coral was believed to be a plant, but we now know that coral branches actually contain living animals called polyps, which resemble little white flowers when open. Coral does not have a root system but rather a flat, disklike base that attaches itself readily to whatever underwater surface may be available.

Because the fragile coral reefs that these communities of tiny sea creatures gradually build up are considered absolutely essential to the world's ecosystems, many areas now forbid the harvesting of living coral reefs. Instead, only specimens that have already broken away from a coral colony are available for harvest, for use in jewelry and decorative items. These restrictions have spawned a wide array of coral simulants. Nevertheless, I have personally seen lots of beautiful, delicate coral wasted in rip-off tourist traps, usually in hideous combinations with neon green seashells or as support devices for thermometers bearing the name of some unimportant landmark.

When it comes to coral jewelry, I have read somewhere that coral can be damaged by excessive perspiration, so I guess you shouldn't wear a coral necklace if you plan on running across Nevada in July. Seriously, coral can react to your body acids and change color, which is why coral jewelry is best in closed settings, with the metal protecting the stone from all angles. Still, on rare occasion you may notice the color of your coral starting to change. If so, a soft polishing cloth, together with a small amount of hydrogen peroxide, may restore its original color.

Red is generally considered the most sought-after variety of coral, but it can be found in a myriad of colors, including pink, white, yellow, and black. These colors often indicate its place of origin. The red, pink, and white varieties are generally associated with the waters off Japan, Africa, and the Mediterranean

Sea, while the more elusive yellow and black specimens are found primarily off the coast of Australia and in the South Pacific. Coral also takes readily to a dye, and as a result it can be found in just about any color you can imagine.

Most of the coral used for jewelry today comes from the Mediterranean Sea. It is usually fashioned into bead form or carved into pins, cameos, and other pendants and brooches. Although coral and turquoise are totally unrelated, the two are sometimes found in combination, inexplicably fashioned into jewelry inspired by the old Southwest. Now, maybe turquoise is common there, but the last time I checked a map of the United States, southwestern hot spots like Phoenix, Las Vegas, and Albuquerque did not have much of a seacoast.

Human use of coral has a long history that is by no means confined to the jewelry trade. In fact, records dating back thousands of years have confirmed its presence in various objects of art and desire. One country that especially seems to like coral is Morocco, where it is the national gemstone. In folklore, coral was believed to prevent ill fortune and, when worn as a necklace, to protect a person from skin disease. To dream about coral signified recovery from a long illness by either the dreamer or a loved one. On the other hand, because it is so sharp, coral can easily cut your bare foot wide open if you step on it; many early fishermen fell victim to its wrath, and therefore some believed it was a sign of the devil. I can attest to its danger, having been carved by coral in Key West; however, I suspect this had more to do with my stupidity than with the devil.

Coral is a nontraditional birthstone for January, April, October, and November. Also significant in the astral world, coral is associated with the signs of Aries, Taurus, and Libra. In ancient times, some observers of the skies believed that Mars was composed of coral, which no doubt accounts for its astral association with the red planet. Coral is also the commemorative gem for people celebrating their thirty-fifth anniversary.

The availability of coral in the jewelry world is certain to undergo even stricter government controls with the passing of time. In fact, as of this writing, coral reefs throughout the world are falling victim to a bacteria that scientists have yet to isolate. If this disturbing situation continues, many of the world's largest coral reefs will be in danger of extinction. Since many colorful species of sea creatures depend on coral for their very lives, it's hoped this bacteria will soon be identified and destroyed, thus protecting the coral reefs and their vital role in the ecosystems of the world. All things considered, I suppose its contribution to the jewelry world is rather insignificant.

Jet

In the simplest terms, jet is a species of fossilized wood that has been immersed for thousands or even millions of years. Like its organic cousin amber, jet is usually mined through cavities drilled deep into immense coastal cliffs and mountains. At times it can also be found along the ocean shore, riding in on the latest tides. It is a very close relative to coal, and it will even burn and smell like coal if it is heated.

The earliest records of jet mining date back to about 1500 B.C., when it was first mined in what is now England. To this day, England remains the most significant source of jet. Others include Poland, France, Spain, and Russia. In the United States, jet can be found in the rugged mountains of Utah, although this source is not reliable. Jet can also be found on the beaches of Florida. In fact, many beachgoers may have passed right by pieces of jet lying undetected in shallow water.

Jet enjoyed heightened popularity during the Victorian era, when fashionable beads, amulets, and hand-carved cameos made of this black organic were in great demand. From that point on, jet jewelry became associated with mourning and death, though historians are not certain about the exact origins of this tradition. Even today, certain cultures believe that dreaming about jet means the onset of great personal sorrow. Astrologers link jet to the planet Saturn.

The market for jet has improved slightly in the past decade, but consumer demand continues to be pretty low. When seen in the jewelry trade today, jet most often is found either in Native American art from the American Southwest or in oriental motifs, often in colorful combinations with jade and coral. Although it takes a high polish and gives off a rich black luster, jet will sometimes dry up, crack, and split.

Because there are so many fine simulants available to consumers these days at a fraction of the cost, natural jet hardly turns heads anymore. If you are looking for black beads and don't wear costume jewelry, the more durable black onyx will probably be a much better choice than jet.

$\mathcal{I}$vory

Ivory dates back to prehistoric time. In fact, archaeological digs have turned up pieces believed to be thirty thousand years old! Cultures in the Orient have prized it not only for articles of adornment and carvings, but for medicinal purposes too. Ivory comes from the teeth or tusks of a variety of animals, the most sought after being the African and Indian elephants. It is also taken from hippos and boars, as well as whales and other sea mammals. Government-imposed restrictions to protect endangered species have narrowed the market for ivory. Many countries, including the United States, now forbid the importation of natural ivory. Sadly, poachers still exist, smuggling ivory into foreign lands in a variety of ways. Ivory takes well to a dye, which sometimes allows it to flow undetected into the marketplace, disguised as jade, coral, or jet. These pieces are almost always seen in bead form. Jasper, bone, vegetable ivory, and fossil ivory are some of the materials currently popular as replacements for ivory from living animals. Craftspeople now incorporate man-made plastic and resin in their statues and other adornments.

Tortoiseshell

Most consumers are surprised to learn that tortoiseshell does not actually come from a tortoise, which lives on land, but rather from a gentle and friendly marine creature of the tropics known as the hawksbill turtle. It was formerly in high demand, as manufacturers crafted tortoiseshell into jewelry, artifacts, eyeglasses, and even writing instruments. At one point, in fact, some cultures considered ownership of tortoiseshell to be a sign of great wealth.

Currently, genuine tortoiseshell, like ivory, is imported and available only on the black market. Conservationists working with various governments have prompted legislation declaring the hawksbill turtle an endangered species. As a result of their efforts, nearly all tortoiseshell seen at retail today is made of a simulated material, generally some form of plastic or resin.

General Guidelines for Care and Cleaning of Organics

All organics are derived from living matter and therefore require special care and cleaning. In general, do not expose any organic matter to harsh detergents, soaps, chemicals, or even hot water. Avoid contact with hair spray, perfume, or powder. Whenever possible, try to handle these pieces by their settings, rather than the gems. Organics are fragile and should be stored in either a pouch or a piece of silk material; they should also be kept in a separate area apart from all other forms of jewelry that could mar their surface. When storing a necklace or a bracelet, it is better to lay it flat for storage than to hang it on a hook.

To clean an organic properly, choose the most conservative measures first. I would start with just a soft polishing cloth, graduating to a rinse in plain water if necessary. If this still doesn't solve your problem, try a few drops of a very mild, detergent-free soap in a warm-water rinse. After drying, you may use a small amount of olive oil to keep your organics moist. If all else fails, consult a trusted jeweler for suggestions. Do not, under any circumstances, try aggressive jewelry cleaners, ultrasonic units, or steam cleaners, because any of them could decompose or permanently alter the color of your treasures. If you follow these guidelines, you will get years of enjoyment from your organic beauties.

One Final Comment

The future of coral is still in the balance. If we abide by the measures that various governments have enacted and aid the efforts of conservationists, coral may survive, even thrive again. Meanwhile, free-floating coral (if you can be sure it didn't float free because of human intervention) is quite suitable for jewelry and artifacts, and that kind will not have an impact on the world's ecosystems.

Thanks to efforts by conservationists and government legislation, ivory and tortoiseshell have pretty much disappeared from conventional retail outlets. Unfortunately, demand lives on across the black markets of the world, and poachers continue to thrive in a dangerous, sometimes deadly, high-stakes profession. Until we as human beings can eliminate the market for natural ivory and tortoiseshell, the future of the helpless creatures those materials come from remains in doubt. There is nothing more tragic than extinction: forever is a long, long time.

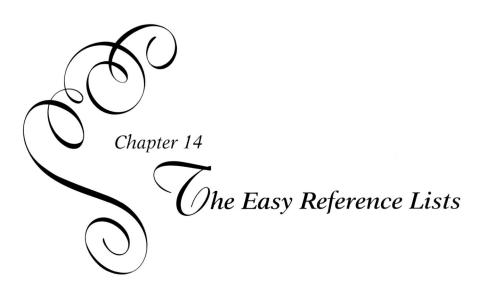

Chapter 14

The Easy Reference Lists

This section of the book contains a series of reference lists that will put important and interesting information right at your fingertips. These lists are a direct result of serious research that I have obtained from well-respected, highly technical professional publications such as the *Sports Illustrated* swimsuit issue and *Mad Magazine*.

Keep in mind that all of these lists are intended to be for your enjoyment, not to establish a central clearinghouse for technical data on gems (although I do include, for example, a list of key properties—hardness, refractive index, and specific gravity—of gems and minerals). If you happen to be an expert in the study of the planets, please don't write long, boring letters telling me that you consider goshenite a stone of Jupiter rather than Venus. Or if science or math is more your specialty, don't bother to send me some heavily researched scientific dissertation that shows the specific gravity of hessonite to be 3.67 rather than 3.65. In all honesty, I couldn't care less. If you have read *Gems: A Lively Guide for the Casual Collector* up to this point, the book's direction should be pretty clear by now. If you're hoping to find lists of chemical composition here, or data on the crystal structure of gems—sorry, pal, I'm afraid you've got the wrong number.

Some of these lists actually result from combinations of information about different cultures and beliefs. Because of this, some may contradict theories that you may subscribe to; in fact, some of the lists may even seem to contradict each other. Traditions, folklore, and interpretations regarding rites of worship for the most part end up being, like opinions, individual matters that we choose to adopt or reject as we travel through this crazy maze we call life. Reject none out of hand; instead, adopt those of particular personal importance, and set aside the others with an open mind for another day.

The Jewelry Shopping Guides

The following bullet lists will provide you with convenient points of reference whether you are shopping for gemstones or pearls. Although the experienced gem shopper may regard some of these points as obvious, I believe they will serve as an excellent reminder to everyone. Since all purchases will have certain things in common, use the general checklist in combination with the more specific lists that follow.

The General Jewelry Shopping Guide

- Make certain that the asking price is clearly indicated, and that it fits within your budget.

- Determine the methods of payment available, including any special financing.

- If you decide to finance your purchase, be sure that the interest rate, terms, and conditions are all in writing and clearly understood.

- Check for appraisals, if any. If none exist, insist that the dealer back the item in writing.

- If this is a major purchase for you, have a written appraisal done *at least* once. *Note: Verbal appraisals are not recognized by insurance companies as a statement of value.*

- Make certain you clearly understand the retailer's return policy, and find out what exceptions, if any, exist. Get this information in writing if at all possible.

- Determine the length and conditions of the guarantee, if any. Whenever possible, obtain this information in writing.

- Make note of the salesperson's name, and request his or her business card.

- Obtain the telephone number and hours of both the local retailer and, if applicable, the company's national headquarters and customer-service department.

- Determine the name and the location of the manufacturer. Request the country of origin whenever possible.

- Where applicable, check to be sure the clasp operates properly and is easy to use. *Note: A clasp that is difficult to operate on the first attempt will usually present a problem later on.*

- Always get a receipt in writing, regardless of the cost of your purchase.

The Gemstone Shopping Guide

- Get a precise identification of the stone.

- Check to see if the gem is natural, synthetic, or a simulant.

- Determine the grade of color and clarity. This is particularly important when shopping for diamonds.

- Find out the carat weight of the individual stone if the piece under consideration is a solitaire, or the total carat weight of the gems if it is a setting with more than one stone.

- Have the metal used in the setting identified. *Note: Make certain it is properly stamped.*

- Examine all of the stones closely to be sure they are not damaged, loose, or improperly set in any manner.

- Check for clarity and the presence of any obvious inclusions in the stones. Look for fractures, cracks, chips, and hazing.

- If the piece contains colored gems, remember to examine them for depth and consistency of color. If the item under consideration is a setting with more than one stone, also make certain the color of every stone matches.

- Examine the setting and the stones from all sides, including the underside, to see if the gems are properly cut for depth and dimension.

- If you own a jewelers loupe, take it along and use it. If not, check with the store: someone there may be able to provide you with one. If nothing else, this will send a positive message on your part to the retailer or dealer.

- Try it on, and check for a snug but proper fit. Make certain it is comfortable, or you may someday regret this purchase.

- Determine the hardness of the gems and ask the retailer if the stones require any special handling and care.

- If the type of stone is considered soft or fragile, stay away from bracelets, because they undergo a lot of abuse even with normal wear. If the piece under consideration is a ring, make certain the gem is well protected by the setting. Even so, generally speaking, stones with a hardness rating of 6.00 or less are best suited for earrings, pendants, or pins.

- If you are at all uncertain, find out if the pearls under consideration are salt-water or freshwater pearls.

- Determine if they are natural or cultured pearls. Since natural pearls are much more expensive than cultured ones, this should be easily apparent to even the most inexperienced consumer.

- Since pearls are measured in millimeters, not by carat weight, check the size of the pearls.

- If applicable, be sure each pearl matches the next in size and luster.

- Run the pearls between your thumb and index finger to see if any of them are rough or gritty. Pearls for the most part are smooth and should never feel like sandpaper.

- Find out how the pearls are strung and ask if they are single- or double-knotted. Double-knotted pearls are preferable.

- If the piece under consideration is a ring, pendant, or pair of earrings, find out about the setting. *Note: Solitaire pearls should be held in place by an internal prong set deep within the body of the pearl. Those held in place by an adhesive are not recommended.*

- If you are buying a strand of pearls, more than likely it will eventually need to be restrung. Determine if the retailer offers this service, and the approximate cost as well. If not, get a recommendation from the retailer.

- If the pearls you are considering are colored, ask if the color is natural, dyed, or irradiated. If possible, find out if the pearls have been bleached to lighten the inner black coating known as conchiolin.

- All pearls are delicate and require special care and cleaning. Ask your retailer for advice, and check to see if any pertinent literature is available.

Birthstones

Birthstones have been around in one form or another nearly as long as people have. Most gem historians believe our ancestors regarded gems as sacred gifts from the heavens. Some people theorize that the earliest associations between gemstones and months of the year evolved from the twelve stones mentioned in the Bible as being on the breastplate of the high priest, while others attribute the development of the birthstones to the Twelve Apostles. Still other religions have produced quite different beliefs.

Needless to say, all of this created quite a controversy as the world of gemstones began to develop. Better means of gem identification, more modern interpretations of various religious writings, and various transformations in the world have all played a pivotal role in the evolution of the birthstone. Finally, in 1912, in an effort to resolve this issue, the American National Retail Jewelers' Association adopted a list of so-called accepted birthstones. (Other groups have issued their own lists since then.) I have designated these gems "traditional birthstones."

However, many different cultures and religions have held to their own beliefs and traditions right up to the present, and I support them wholeheartedly. Personally, I think that too often in the hustle and bustle of our modern world, we put aside traditions and compromise our beliefs just so we can fit in with the rest of society. This in itself is a shame.

Therefore, I have decided to honor every alternative birthstone I could find, in the hope of keeping all the traditions alive. You will find some gems listed several times; though this seems to be a contradiction, remember that the various different lists that inspired this one alternative list evolved over many hundreds or thousands of years. Only for the sake of reference, the gems on this second list are classified as "nontraditional birthstones." This is in no way intended to relegate these gems to a secondary status. I feel that when a governing body decides to establish a tradition, it shouldn't be at the expense of other people's beliefs. Feel free to accept the ones you want.

Traditional Birthstones

January	Garnet (Almandine)
February	Amethyst
March	Bloodstone
April	Diamond
May	Emerald
June	Pearl
July	Ruby
August	Sardonyx
September	Sapphire
October	White Precious Opal
November	Imperial Topaz
December	Turquoise

Nontraditional Birthstones

(Listed in alphabetical order for each month)

January	Coral, Emerald, Hyacinth, Rose Quartz, Sapphire, and Serpentine
February	Diamond, Garnet, Hyacinth, Kunzite, Onyx, Pearl, and Topaz
March	Aquamarine, Hematite, Hyacinth, Jasper, and Topaz
April	Coral, Moonstone, Pearl, Quartz Crystal, White Sapphire, and Zircon
May	Agate, Carnelian, Chrysoprase, Garnet, Jade, and Tourmaline
June	Agate, Alexandrite, Emerald, Moonstone, Sapphire, and Turquoise
July	Carnelian, Diamond, Green Sapphire, Onyx, Pearl, Sardonyx, and Turquoise
August	Alexandrite, Aventurine Quartz, Carnelian, Emerald, Green Topaz, Moonstone, Peridot, and Zircon
September	Blue Zircon, Emerald, Lapis Lazuli, Peridot, Sardonyx, and Topaz
October	Aquamarine, Coral, Garnet, Diamond, Morganite, Moonstone, Pink Sapphire, Pink Topaz, Pink Tourmaline, and Pink Zircon
November	Amber, Citrine, Coral, Heliodor, Pearl, Peridot, and Sapphire
December	Bloodstone, Blue Apatite, Blue Topaz, Blue Zircon, Chrysoprase, Hematite, Lapis Lazuli, Ruby, and Star Sapphire

Astral Stones

Our ancestors attributed a wide variety of magical powers to colored gemstones. Early physicians believed that jade could cool the burning forehead of a child sick with fever. Christians believed that when amber was found, it signified the presence of the Lord. Certain cultures regarded the black organic jet as a sign of great sorrow to come. Ancient Egyptians buried their dead with a scarab of lapis lazuli, because they felt it would protect the deceased on their journey into the afterlife. The list goes on forever.

Because many of these apparent attributes could not be explained in any material manner, astrologers turned to the heavens for answers. As in the case of birthstones, astral stones varied from culture to culture and religion to religion. In order to simplify things just a bit, I have broken these astral stones down into two separate groups: traditional and nontraditional ones.

No central authority has ever developed a single traditional list, as the American National Retail Jewelers' Association did with birthstones in 1912, but certain traditions are more dominant than others. I have chosen the beliefs of the early Hindus as the "traditional" source of these astral stones. In my research, I have found their influence to be quite strong and widely accepted. As with birthstones, I created a second category of so-called nontraditional astral stones, so that all the beliefs and traditions I have happened upon will be honored and kept alive. You should feel free to reclassify any or all of the astral signs according to their significance in your life.

Incidentally, I also found that the dates of the astral signs vary from one tradition to another, just about as often as the astral stones do themselves. This is why you may come across a conflict between the dates in this book and those in your local newspaper or magazine. Remember, all of this is based on tradition, cultural belief, and hearsay, so take from it what you may, in the spirit of fun in which it's presented.

Traditional Astral Stones

Aquarius *(January 22–February 21)*	Garnet
Pisces *(February 22–March 21)*	Amethyst
Aries *(March 22–April 20)*	Bloodstone
Taurus *(April 21–May 21)*	Sapphire
Gemini *(May 22–June 21)*	Agate
Cancer *(June 22–July 22)*	Emerald
Leo *(July 23–August 22)*	Onyx
Virgo *(August 23–September 22)*	Carnelian
Libra *(September 23–October 23)*	Peridot
Scorpio *(October 24–November 21)*	Aquamarine
Sagittarius *(November 22–December 21)*	Imperial Topaz
Capricorn *(December 22–January 21)*	Ruby

Nontraditional Astral Stones

(Listed in alphabetical order for each astral sign)

Aquarius *(January 22–February 21)*	Amber, Amethyst, Aquamarine, Blue Zircon, Hematite, Lapis Lazuli, and Turquoise
Pisces *(February 22–March 21)*	Amber, Bloodstone, Cat's-Eye, Jade, Peridot, and Quartz Crystal
Aries *(March 22–April 20)*	Amethyst, Carnelian, Coral, Malachite, Rose Quartz, and Sard
Taurus *(April 21–May 21)*	Agate, Azurite, Carnelian, Coral, Emerald, and Lapis Lazuli
Gemini *(May 22–June 21)*	Alexandrite, Citrine, Emerald, Pearl, Peridot, and Tiger's-Eye
Cancer *(June 22–July 22)*	Amber, Diamond, Moonstone, Onyx, Pearl, and Ruby
Leo *(July 23–August 22)*	Amber, Citrine, Garnet, Hyacinth, Peridot, Quartz Crystal, and Ruby
Virgo *(August 23–September 22)*	Agate, Peridot, Rhodochrosite, Ruby, Sapphire, and Sardonyx
Libra *(September 23–October 23)*	Agate, Coral, Diamond, Jade, Sapphire, Smoky Quartz, and White Precious Opal
Scorpio *(October 24–November 21)*	Amber, Bloodstone, Jasper, Malachite, Ruby, Sard, Sardonyx, and White Precious Opal
Sagittarius *(November 22–December 21)*	Blue Zircon, Jasper, Malachite, Moonstone, and Tourmaline
Capricorn *(December 22–January 21)*	Agate, Amethyst, Cat's-Eye, Garnet, Onyx, Pyrope, Quartz Crystal, Turquoise, and White Precious Opal

The Chinese Zodiac

The Chinese traditionally observe specific character traits based on the year of a person's birth, following a twelve-year cycle of signs named for the twelve animals listed below. Since not many of you are more than one hundred years old, this list stops at the year 1898. As time goes on, keep track of your particular sign by adding twelve to the most recent year indicated. Enjoy!

The Year of the Tiger
    1998, 1986, 1974, 1962, 1950, 1938, 1926, 1914, 1902
    Famous fellow tigers: Dwight Eisenhower, Marilyn Monroe, Natalie Wood
The Year of the Ox
    1997, 1985, 1973, 1961, 1949, 1937, 1925, 1913, 1901
    Famous fellow oxen: Princess Diana, Jack Nicholson, Dan Dennis
The Year of the Rat
    1996, 1984, 1972, 1960, 1948, 1936, 1924, 1912, 1900
    Famous fellow rats: Marlon Brando, Gene Kelly, William Shakespeare
The Year of the Pig
    1995, 1983, 1971, 1959, 1947, 1935, 1923, 1911, 1899
    Famous fellow pigs: Fred Astaire, Ernest Hemingway, Alfred Hitchcock
The Year of the Dog
    1994, 1982, 1970, 1958, 1946, 1934, 1922, 1910, 1898
    Famous fellow dogs: Judy Garland, Michael Jackson, Sylvester Stallone
The Year of the Rooster
    1993, 1981, 1969, 1957, 1945, 1933, 1921, 1909
    Famous fellow roosters: John Glenn, Goldie Hawn, Carly Simon
The Year of the Monkey
    1992, 1980, 1968, 1956, 1944, 1932, 1920, 1908
    Famous fellow monkeys: Bette Davis, Nelson Rockefeller, Elizabeth Taylor
The Year of the Sheep
    1991, 1979, 1967, 1955, 1943, 1931, 1919, 1907
    Famous fellow sheep: Robert De Niro, Barbara Walters, John Wayne
The Year of the Horse
    1990, 1978, 1966, 1954, 1942, 1930, 1918, 1906
    Famous fellow horses: Pearl Bailey, Sean Connery, Rita Hayworth
The Year of the Snake
    1989, 1977, 1965, 1953, 1941, 1929, 1917, 1905
    Famous fellow snakes: Henry Fonda, John F. Kennedy, Mae West
The Year of the Dragon
    1988, 1976, 1964, 1952, 1940, 1928, 1916, 1904
    Famous fellow dragons: Shirley Temple Black, John Lennon, Ringo Starr
The Year of the Rabbit
    1987, 1975, 1963, 1951, 1939, 1927, 1915, 1903
    Famous fellow rabbits: Bob Hope, Orson Welles, Linda Dennis

The Planetary Influence

Our earliest ancestors were fascinated with the sun and the moon possibly as far back as the beginning of humankind. As science began to develop and other planets were discovered, they became a source of both worship and great fear. The desire to learn more about these distant bodies burns brightly within us even today.

At first, people knew very little about gemstones, and even less about the planets. This provided for some interesting folklore that linked the two together. Moonstone was considered a direct link to the moon; in fact, some even felt the moonstone was actually made of pebbles from the moon. Whenever the moon would go into an eclipse, it inspired terror among members of early civilizations. Consequently, moonstone was blamed for the disappearance of the moonlight.

Early documents also provide evidence about early cultures' fear of the sun. They considered the sun to be all-powerful. When temperatures soared, they thought the sun was in a rage. Since the ruby was considered the most powerful gemstone of all, it was felt to signify the anger of the sun. In some Far Eastern cultures, the hyacinth, or golden zircon, was linked to the dragon that was believed to control the eclipse of the sun and the moon.

As time went on and additional planets were discovered, astrologers continued to promote the idea of a relationship between gemstones and the planets. Some believed the planet Mars was made of coral. The chalcedony known as sardonyx, with its multicolored bands, was thought to have originated on Saturn. Cat's-eye gems so baffled astrologers that they attributed those stones to Ketu, the southern lunar node (where the sun and moon appear to cross paths in the sky), a site where unexplained phenomena are believed to develop.

As in the case of birthstones and astral stones, the stones of the planets are mostly conjecture in combination with a heavy dose of tradition and religion. The list that follows takes into account every single relationship I could find between gemstones and their planets. I hope you enjoy them!

Stones of the Planets

(Listed in alphabetical order for each planet and other astronomical body or site)

Jupiter	Chalcedony, Citrine, Heliodor, Hyacinth, Imperial Topaz, and Yellow Sapphire
Ketu*	Cat's-Eye Apatite, Cat's-Eye Chrysoberyl, and Cat's-Eye Tourmaline
Mars	Bloodstone, Carnelian, Coral, Hematite, Hyacinth, Pyrope, Ruby, and Sardonyx
Mercury	Agate, Chrysoprase, Emerald, Green Tourmaline, Hematite, Jade, Jasper, Peridot, Topaz, Tsavorite, and Yellow Sapphire
Moon	Emerald, Hyacinth, Moonstone, and Pearl
Rahu*	Hessonite, Red-Orange Zircon, and Spessartine
Saturn	Amethyst, Blue Sapphire, Blue Spinel, Chalcedony, Iolite, Jasper, Jet, Lapis Lazuli, Sardonyx, and Tanzanite
Sun	Chrysoberyl, Diamond, Hyacinth, Pyrope, Red Spinel, Rubellite, and Ruby
Venus	Chrysoprase, Diamond, Emerald, Goshenite, Jasper, Lapis Lazuli, Peridot, Quartz Crystal, Topaz, Turquoise, White Sapphire, and Zircon

*Rahu and Ketu are not actually planets, of course, but rather lunar nodes, the places in the celestial sphere where the sun and moon appear to cross paths (resulting in eclipses and, some believe, various unexplained phenomena). Rahu is the northern lunar node; Ketu, the southern lunar node.

Commemorative Stones

After endless years of research, I have established a theory that every gem is linked in one way or another to some sort of a list so that people will feel a certain obligation to own all of them. I suspect that this is one giant conspiracy, put together by the gem people, the flower people, the candy people, and the greeting-card people. Oh, I could throw in a couple more, like the stuffed-animal people and the restaurant people, but even without them I think you get the idea. There are lists that link gemstones to everything from the seasons of the year to a person's name. If you know what time and day of the week you were born, other purchasing opportunities are ready and waiting!

What follows next is the list of anniversary stones, in order of year. You will see at a glance that certain anniversaries have more than one stone associated with them. This usually means that at least one religion, custom, or tradition contradicts another. Lest anyone be offended, I have arranged these gems in alphabetical order for each year. As for which cultures and traditions inspired which of the commemorative links listed below, well, if you can come up with any, you're much better than I am.

Anniversary Stones

1. Gold
2. Garnet
3. Pearl
4. Blue Topaz
5. Sapphire
6. Amethyst
7. Onyx
8. Tourmaline
9. Lapis Lazuli
10. Amber and Diamond
11. Turquoise
12. Agate and Jade
13. Citrine and Moonstone
14. Agate and Opal
15. Quartz Crystal and Ruby
16. Imperial Topaz and Peridot
17. Amethyst
18. Garnet
19. Aquamarine and Hyacinth (Golden Zircon)
20. Emerald
23. Blue Sapphire and Blue Topaz
25. Silver
26. Blue Star Sapphire
30. Pearl
35. Coral and Emerald
39. Cat's-Eye
40. Ruby
45. Alexandrite and Sapphire
50. Gold
52. Star Ruby
55. Alexandrite and Emerald
60. Diamond

$\mathcal{F}$inding Gems and Minerals in Your Own Backyard

When you think of rock and mineral deposits in the United States, if you are like most of us, certain states come to mind. Arizona, California, Maine, Montana, and the Carolinas all have well-established reputations as being rich in gems. But you might be surprised to learn that star garnets can be found in Idaho, and that New Jersey contains important deposits of rhodonite and smoky quartz. It's true! In fact, every state contains deposits of at least one gem or mineral.

What follows is a list of gems and minerals that can be found far and wide across our great nation (including many that otherwise have no place in this book on gems). I have compiled this list to the best of my ability, but I still suspect it is far from complete. It isn't meant to be the world's foremost authority on geology, but rather a site of easy reference and a source of local information. If you are so inclined, investigate a little further. Heck, for all you know, if you live in Colorado, rose quartz could surface in your own backyard!

Gems and Minerals of the United States

(Listed in alphabetical order)

State	Gems and Minerals
Alabama	Agate, Aquamarine, Cacoxenite, Heliodor, and Hematite
Alaska	Almandine Garnet and Jade
Arizona	Azurite, Caledonite, Chalcedony, Copper, Cuprite, Fire Agate, Garnet, Malachite, Nephrite Jade, Peridot, and Turquoise
Arkansas	Agate, Cacoxenite, Diamond, Smithsonite, Smoky Quartz, and Wavellite
California	Axinite, Benitoite, Blue Chalcedony, Chrysoprase, Cinnabar, Colemanite, Diamond, Elbaite, Garnets, Goldenite, Goshenite, Graphite, Hiddenite, Jadeite Jade, Jasper, Kunzite, Lapis Lazuli, Lazulite, Morganite, Nephrite Jade, Powellite, Tantalite, Titanite, Tourmaline, and Turquoise
Colorado	Agate, Amethyst, Aquamarine, Bismuthinite, Diamond, Fluorite, Garnet, Golden Quartz, Lepidolite, Microline, Pyrite, Rhodochrosite, Rose Quartz, Siderite, Sylvanite, and Turquoise
Connecticut	Danburite, Garnet, Rose Quartz, Tourmaline, and Wurtzite
Delaware	Feldspar and Sillimanite
Florida	Chalcedony, Coral, and Jet
Georgia	Amethyst, Garnet, Lazulite, Quartz Crystal, Ruby, and Tektite

Hawaii	Pearl, Peridot, and Obsidian
Idaho	Agate, Garnet, Ilvaite, Ludlamite, Obsidian, Opal, Sapphire, Sillimanite, and Vivianite
Illinois	Alstonite, Fluorite, Sphalerite, Strontianite, and Witherite
Indiana	Agate and Chalcedony
Iowa	Agate, Chalcedony, and Quartz
Kansas	Calcite and Chalcedony
Kentucky	Fortification Agate and Millerite
Louisiana	Agate and Feldspar
Maine	Beryl, Beryllonite, Garnet, Goshenite, Hessonite, Petalite, Quartz Crystal, Schorl, Topaz, and Tourmaline
Maryland	Beryl, Chalcedony, and Quartz
Massachusetts	Diaspore, Goshenite, Jasper, Margarite, Parisite, and Rhodonite
Michigan	Agate, Chalcedony, Greenstone, Powellite, and Pumpellyite
Minnesota	Chalcedony, Greenstone, and Prehnite
Mississippi	Chalcedony and Freshwater Pearl
Missouri	Calcite, Chalcopyrite, Dolomite, Fluorite, Leadhillite, Pyrite, and Siegenite
Montana	Agate, Amethyst, Calcite, Chalcedony, Garnet, Kyanite, Parisite, Pyrite, Obsidian, Quartz Crystal, Sapphire, and Wurtzite
Nebraska	Blue Agate and Chalcedony
Nevada	Axinite, Chalcedony, Clinoclase, Goethite, Goldenite, Nephrite Jade, Powellite, Quartz Crystal, Realgar, Rhyolite, Spessartine, and Turquoise
New Hampshire	Augelite, Brazilianite, Calcite, Chalcedony, Fluorite, Graftonite, Gummite, Lazulite, Schorl, Smoky Quartz, and Titanite
New Jersey	Agate, Augite, Fluorite, Franklinite, Lazulite, Prehnite, Rhodonite, Smoky Quartz, Sussexite, and Zincite
New Mexico	Calcite, Chalcopryite, Chrysocolla, Garnet, Jade, Malachite, Obsidian, Peridot, Quartz Crystal, Smithsonite, Tantalite, and Turquoise
New York	Artinite, Brewsterite, Chondrodite, Danburite, Diopside, Fluorite, Garnet, Graphite, Herkimer Quartz, Magnetite, Millerite, Rose Quartz, Strontianite, Tourmaline, Uvite, and Witherite
North Carolina	Aquamarine, Chalcedony, Diamond, Garnet, Hiddenite, Hypersthene, Kunzite, Kyanite, Quartz Crystal, Rhodolite, Ruby, Samarskite, Sapphire, Thulite, Torbernite, and Tourmaline

North Dakota	Agate, Chalcedony, and Quartz
Ohio	Chalcedony and Feldspar
Oklahoma	Agate, Malachite, Quartz Crystal, and Turquoise
Oregon	Agate, Blue Chalcedony, Goldenite, Hornblende, Hypersthene, Moonstone, Obsidian, Opal, Sunstone, and Tridymite
Pennsylvania	Amethyst, Garnet, Millerite, Moonstone, Serpentine, Strontianite, Sunstone, and Titanite
Rhode Island	Bowenite and Quartz Crystal
South Carolina	Amethyst, Garnet, Quartz Crystal, Ruby, and Sapphire
South Dakota	Agate, Barite, Calcite, Chalcedony, Graftonite, and Rose Quartz
Tennessee	Agate, Chalcedony, Freshwater Pearl, Quartz Crystal, and Zoisite
Texas	Chalcedony, Diamond, Gadolinite, Jade, Tektite, Topaz, Tourmaline, and Turquoise
Utah	Austinite, Azurite, Chalcedony, Garnet, Malachite, Red Beryl, Smithsonite, Tridymite, Turquoise, Variscite, and Wardite
Vermont	Beryl, Chalcedony, Goethite, Magnetite, and Schorl
Virginia	Amethyst, Garnet, Moonstone, Quartz Crystal, and Spessartine
Washington	Agate, Autunite, Chalcedony, Obsidian, and Realgar
West Virginia	Amethyst, Garnet, and Quartz Crystal
Wisconsin	Agate and Chalcedony
Wyoming	Agate, Chalcedony, Jade, Shortite, and Trona

$\mathcal{K}$ey Properties of Gems and Minerals

(Listed in alphabetical order)

Gem or Mineral	Hardness	Refractive Index	Specific Gravity
Agate	7.00	1.53–1.54	2.61
Alexandrite	8.50	1.74–1.75	3.71
Almandine Garnet	7.50	1.76–1.83	4.00
Amber*	2.50	1.54–1.55	1.08
Amethyst	7.00	1.54–1.55	2.65
Ametrine	7.00	1.54–1.55	2.65
Andalusite	7.50	1.63–1.64	3.12
Andradite	6.50	1.85–1.89	3.85
Apatite	5.00	1.63–1.64	3.20
Aquamarine	7.50	1.57–1.58	2.69
Aventurine Quartz	7.00	1.54–1.55	2.65
Azurite	3.50	1.73–1.84	3.78
Black Opal	6.00	1.37–1.47	2.10
Bloodstone	7.00	1.53–1.54	2.60
Blue Chalcedony	7.00	1.53–1.54	2.61
Blue Sapphire	9.00	1.76–1.77	4.00
Blue Topaz	8.00	1.62–1.63	3.54
Calcite	3.00	1.48–1.66	2.71
Carnelian	7.00	1.53–1.54	2.60
Cat's-Eye Chrysoberyl	8.50	1.74–1.75	3.71
Cat's-Eye Quartz	7.00	1.54–1.55	2.65
Chrysoprase	7.00	1.53–1.54	2.61
Citrine	7.00	1.54–1.55	2.65
Coral*	3.50	1.49–1.66	2.65
Danburite	7.50	1.63–1.64	3.00
Demantoid	6.50	1.85–1.89	3.85
Diamond	10.00	2.42	3.52
Diopside	5.50	1.66–1.72	3.29
Dioptase	5.00	1.67–1.72	3.31
Emerald	7.50	1.57–1.58	2.70
Fire Opal	6.50	1.45	2.20
Fluorite	4.00	1.43	3.10
Goldenite	7.00	1.53–1.54	2.65
Goshenite	7.50	1.58–1.59	2.80
Grossular Garnet	7.00	1.69–1.73	3.49
Heliodor	7.50	1.57–1.58	2.80

Hematite	6.50	2.94–3.22	5.20
Hessonite	7.25	1.73–1.75	3.65
Hiddenite	7.00	1.66–1.67	3.18
Hyacinth	7.50	1.93–1.98	4.70
Indicolite	7.50	1.62–1.68	3.06
Iolite	7.00	1.53–1.55	2.63
Ivory*	2.50	1.53–1.54	1.90
Jadeite Jade	7.00	1.66–1.68	3.33
Jasper	7.00	1.53–1.54	2.61
Jet*	2.50	1.64–1.68	1.38
Kunzite	7.00	1.66–1.67	3.17
Kyanite**	7.00/5.00	1.71–1.73	3.60
Labradorite	6.00	1.56–1.57	2.70
Lapis Lazuli	5.50	1.50	2.80
Mabe Shell*	2.50	1.53–1.69	1.30
Malachite	4.00	1.85	3.80
Mandarin Garnet	7.00	1.79–1.81	4.16
Milarite	5.50	1.53–1.55	2.61
Moldavite	5.00	1.48–1.51	2.40
Moonstone	6.00	1.52–1.53	2.57
Morganite	7.50	1.58–1.59	2.80
Nephrite Jade	6.50	1.61–1.63	2.96
Obsidian	5.00	1.48–1.51	2.35
Onyx	7.00	1.53–1.54	2.60
Opal	6.00	1.37–1.47	2.10
Orthoclase	6.00	1.51–1.54	2.56
Padparadscha	9.00	1.76–1.77	4.00
Pearl*	3.00	1.53–1.68	2.70
Peridot	6.50	1.64–1.69	3.34
Petrified Wood	7.00	1.53–1.54	2.61
Pyrope	7.25	1.72–1.76	3.80
Quartz Crystal	7.00	1.54–1.55	2.65
Red Beryl	7.50	1.57–1.58	2.80
Rhodochrosite	4.00	1.60–1.80	3.70
Rhodolite	7.00	1.72–1.76	3.85
Rhodonite	6.00	1.71–1.73	3.60
Rose Quartz	7.00	1.54–1.55	2.65
Rubellite	7.50	1.62–1.64	3.06
Ruby	9.00	1.76–1.77	4.00
Sapphire	9.00	1.76–1.77	4.00
Sard	7.00	1.53–1.54	2.61
Sardonyx	7.00	1.53–1.54	2.61

Scapolite	6.00	1.54–1.58	2.70
Serpentine	2.50–5.00	1.55–1.56	2.60
Sillimanite	7.50	1.66–1.68	3.25
Smoky Quartz	7.00	1.54–1.55	2.65
Sodalite	5.50	1.48	2.27
Spectrolite	6.00	1.56–1.57	2.70
Spessartine	7.00	1.79–1.81	4.16
Spinel	8.00	1.71–1.73	3.60
Sunstone	6.00	1.54–1.55	2.65
Tanzanite	6.50	1.69–1.70	3.35
Tiger's-Eye Quartz	7.00	1.54–1.55	2.65
Titanite	5.00	1.84–2.03	3.53
Topaz	8.00	1.62–1.63	3.54
Tortoiseshell*	2.50	1.53–1.69	1.30
Tourmaline	7.50	1.62–1.64	3.06
Tsavorite	7.00	1.69–1.73	3.49
Turquoise	6.00	1.61–1.65	2.80
Uvarovite Garnet	7.50	1.86–1.87	3.77
Zircon	7.50	1.93–1.98	4.69

*Not technically a gem or mineral, but rather organic matter.

**Kyanite has two hardness values: 7.00 when crystals are perpendicular to its cleavage, and 5.00 when they are parallel.

General Guidelines for Care and Cleaning

This section of the book is designed to help you care for and clean your treasures to keep them looking brand-new. First off, remember to consider not only the gemstone, but the type of metal it is set in as well. Generally speaking, this is more important with an overlay product than it is with either a karat-gold, silver, or platinum setting. Please keep in mind that these are only broad guidelines; the final decision is up to you, based in part on more specific advice from your retailer. I assume no legal responsibility whatsoever for damage that may occur in the care or cleaning process.

Conservative Measures Come First

It is wise to take the time to clean your beautiful little treasures as often as possible, because that will pay big dividends in the long run. Always start with the most conservative method possible, which should be a soft polishing cloth. If this does not produce satisfactory results, try filling a bowl with a warm-water solution that contains a few drops of a mild dish-cleaning soap; be sure to avoid harsh chemicals such as ammonia. Soak the object in the solution and then clean with a *soft* brush, paying special attention to the sides and underparts of the stone. Next, rinse with warm, ordinary tap water, and air-dry or pat with a soft cloth.

In most cases, this method will prove satisfactory. If you are still not pleased, purchase a jar of commercial cleaning solution. I have found this cleaner readily available at department stores and drugstores as well as retail jewelers. You should not pay more than about $10 for a bottle. When you purchase a bottle of cleaning solution, you will find that the brush usually comes with it, as part of the plastic dip tray inside. Follow the manufacturer's directions on the bottle or the outer box, if any. For the most part, these solutions are safe; however, if the one you select contains ammonia, you do need to be more careful with it. Check the bottle for the ingredients, or simply ask the salesperson for help.

Ultrasonic Cleaners

Before using an ultrasonic cleaner, you would be wise to warm your jewelry first in a plain-water dip, because great changes in temperature might result in damage to the piece. Ultrasonic cleaners operate on wavelengths that are conveyed through the water or cleaning solution in the dip tray. You will find that the dip tray suspends the piece in the solution so that the jewelry does not touch the bottom of the compartment. This is by design, since such contact could damage your jewelry. You should make certain the setting and stone, or stones, are secure before using this method.

Steam Cleaners

Consumers seldom use steam cleaners at home, but in some cases they are the best choice to remove very stubborn, unsightly marks. Rather than resort to doing it yourself, you should find a jeweler you have confidence in and have him or her evaluate the piece; the jeweler may have a steam cleaner available in the store. If I were you, I would let the expert decide whether steam cleaning is even an option in each particular case. Also keep in mind that any gem displaying color-change properties (such as the alexandrite) can lose this effect immediately if it comes in contact with very hot water. If you choose to use this method yourself, remember to place the gem into a bowl of warm, detergent-free water to gradually lower its temperature after cleaning.

Boiling

Another way to keep some gems clean is right there in your kitchen: boil them! First, mix a mild solution of water and gentle soap or detergent, being certain that the article of jewelry is in place before boiling. Always start out at roon temperature, suspending the jewelry by using a strainer or some other device that will allow the water to flow through all the nooks and crannies of the piece. Bring the mixture to a boil for about ten minutes. Allow it to cool, then rinse with warm (but mild) tap water. Always be certain that you don't expose the article to any drastic changes in temperature. In my opinion, this still seems to be a rather radical method of cleaning, which should be done only on certain gems after the more conservative methods are tried. Diamonds, rubies, sapphires, topaz, jade, and quartz have proved in the past to stand up under this method of cleaning. Gems such as emeralds, opals, tanzanites, and any color-change stones (such as alexandrites) should definitely *not* be boiled.

Special Cases

Remember, there are certain gems that simply should not be exposed to chemicals, ultrasonics, or other aggressive measures of cleaning. Pearls, for example, should *never* be soaked, because this could damage or discolor the pearls or eventually weaken the silk thread that keeps them together. In fact, it would be a very good idea to clean any organic—amber, coral, ivory, jet, pearls, and tortoiseshell—with the utmost of care.

You should also clean opals using the most conservative means that are effective. Since opals can be as much as 10 percent water, do not expose them to heat, since this could cause them to dry up and crack. It is a good idea to keep them moist, with a plain-water mist. When storing opals for a long period of time, it is best to wrap them in cheesecloth soaked with glycerin.

Things I Would Not Do

What follows is a list of things that I definitely would avoid during the care and cleaning of gems and jewelry. Please keep in mind that these are only guidelines, designed to protect and preserve the beauty of your jewelry for life.

- *Never,* under any circumstances, use abrasives or harsh chemicals to clean your jewelry. *Note: Most toothpaste contains abrasives.*
- Use only a minimal amount of soap in making a jewelry dip.
- As a rule, don't use ammonia unless it is absolutely necessary. When cleaning more durable stones such as diamonds, rubies, sapphires, cubic zirconia, and most species of quartz, a few drops of ammonia in a bowl of warm water will be effective. But remember that certain more fragile gems (such as emeralds and opals) can be damaged by ammonia or another aggressive cleaning solution. If you are at all uncertain, check with a jeweler before you use any, because every circumstance is not the same.
- Do not expose treated gems to either an ultrasonic or a steam cleaner.
- Organics and fragile stones should always be kept in their own individual compartment or pouch for protection.
- Do not use a hair dryer to dry your gems. This is just another form of heat, which can discolor or otherwise damage your jewelry.
- Do not wear fragile gems on your wrist or pinkie, for this is where they are most likely to be hit and possibly damaged.
- Do not use bleach—ever!
- Always remember that swimming pools contain chlorine, a harsh chemical that can be the road to ruin for your beautiful gems and jewelry.
- Speaking of the pool, some gems can dry up or fade with prolonged exposure to sunlight. It is usually best to wear costume jewelry in these instances.
- Use radical cleaning methods (abrasives, harsh chemicals, heat) *only* after you have decided the item is no longer wearable. If your only other choice is to discard it, you may as well give it your best shot.

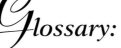lossary:
Understanding the Language of Gemology

Adularescence: The white or silver-blue haze displayed by top-quality moonstones and certain other gems.

Akoya oysters: Mollusks of choice in the cultivation of saltwater pearls.

Alluvial deposits: Gem deposits found in water after they have been separated from the mother rock.

Appraisal: An evaluation of the actual replacement value of a piece of jewelry. This practice is often done by a licensed gemologist.

Asterism: The star effect displayed by certain gems with properly intersecting inclusions. Gems of this variety are always cut *en cabochon* (*see* Cabochon cut).

Baroque: Odd-shaped pearls formed during cultivation.

Boiling: A radical-sounding way to clean certain gems—but definitely not others. Boiling should be done very carefully and only after trying more conservative methods.

Brilliant cut: Round-shaped stone consisting of a minimum of fifty-eight facets.

Cabochon cut: The art of rounding a gem without facets into the shape of a highly polished dome.

Cameo: The art of carving a shell or similar matter above its background.

Carat: Used to denote the weight of gems.

Chatoyancy: The ability of certain gems to display a "cat's-eye" effect, due to the precise occurrence of narrow inclusions. These gems are always cut *en cabochon*.

Cleavage: Breakpoint or weakness of a gem, connected to its atomic structure. Ironically, gems defined as having perfect cleavage are the ones most likely to break when being cut or faceted.

Color: Important property used in the evaluation of a gem, particularly the diamond. Interestingly enough, the quality of a gem may be measured based on either the presence or the absence of color.

Commercial jewelry cleaner: A solution that usually incorporates water and a mild detergent, although many often contain small amounts of ammonia. These solutions are formulated to maintain a proper pH balance (between too much acidity and too much alkalinity).

Conchiolin: Dark substance a mollusk secretes for protection during cultivation.

Crown: Denotes the top part of the gemstone.

Culet: The lowest part of a gem.

Cultivation: The process of forming a pearl by inserting tissue from a sacrifice mollusk into another.

Cut: The method of faceting a gem.

Dichroism: The ability of certain gems to display a second color when viewed from a different angle.

Dispersion: The splitting of light as it enters a gemstone. Also known as the stone's "fire."

Doublet: A stone made of two components, generally held together with a clear or colored adhesive.

Emerald cut: *See* Step cut.

Enhancement of gems: The process by which the appearance of a gem is improved. An excellent example of this is heat enhancement, which most often is used to clarify a gem or improve its color.

Facet: The cut and polished portion of a gemstone.

Faceting: The art of cutting a stone or precious metal to enhance its appearance.

Fire: The rainbow of colors resulting from light rays as they enter a stone.

Fluorescence: The ability of certain gems to glow when heated in low light.

Foiling: The practice of inserting a colored or silver foil behind a gem in a closed-end setting, to enhance its appeal. Most often used in the production of rhinestones.

Full cut: A round-shaped, brilliant-cut gem. Although this is most often used to describe accent diamonds, the term also may apply to colored gems.

Gemstone: A mineral or combination of minerals displaying a high degree of beauty, rarity, durability, and desirability.

Geode: A hollow rock cavity, which usually contains some form of one or more gems. Two excellent examples of gems often found in geodes are amethyst and peridot.

GIA: The Gemological Institute of America, which most experts consider the established authority on gemology. Among many other things, the GIA offers in-home study courses that cover an extensive range of subjects.

Girdle: The widest point (in circumference) of a gemstone.

Gram: Unit of measurement of weight (with just over twenty-eight grams equaling one ounce), often expressed in milligrams (at one thousand milligrams to a gram); in jewelry, grams are used in expressing the weight of precious metals such as gold and silver.

Handmade: A legal term used to describe an item made without the use of machine-driven tools.

Heat treatment: The application of heat to a gem for the purpose of improving its color or clarity. Many gems are treated in this manner.

Illusion setting: Any setting that is not as it appears. Most often, illusion settings combine one or more small diamonds with rhodium to enhance the diamonds, or to give the impression of a higher carat weight.

Inclusions: Foreign matter, within a gem or mineral, that often displays different colors from the rest of the stone, or other effects.

Intaglio: The art of carving a shell or similar matter beneath its background. An intaglio is the opposite of a cameo.

Iridescence: The play of colors seen within a gemstone as a result of inclusions interfering with light entering the stone.

Karat: Unit of measurement that indicates the quantity (or parts) of fine gold in a piece of jewelry. This karatage is always based on pure 24-karat gold. As an example, 18-karat gold contains eighteen parts fine gold and six parts other metal alloys.

Lapidary: Gifted craftsperson who cuts and polishes gems to their finished state.

Lava: Molten rock that forms above the surface of the Earth as a result of volcanic activity.

Loupe: A compact magnifying glass used to examine stones and settings. The most common jewelers loupe magnifies objects ten times.

Luster: The outward appearance of a gem or an organic material. Luster is of particular importance in determining the quality of a pearl.

Mabe: Literally translated, mabe means "half," and this term is used to denote the large half pearl or shell often seen in jewelry. Also known as blister pearls.

Magma: Rock that forms below the surface of the Earth as a result of volcanic activity.

Mineral: An inorganic element of the Earth of consistent atomic structure and chemical composition.

Mohs' hardness scale: Numerical scale developed in the nineteenth century by Friedrich Mohs that assigns a rating to a gem according to its ability to resist scratching (with hardest rated 10 and softest, 1).

Mother-of-pearl: The actual substance that lines the inside of the oyster or other mollusk.

Nacre: The substance secreted by the oyster or other mollusk that surrounds the darker conchiolin and subsequently forms the outer layers of the pearl.

Natural: Gem formed without the assistance of humans.

Opalescence: A variety of iridescence that is most often light blue in color.

Organic gem: Matter that is not technically a gemstone, but rather is derived from animal or plant life. In jewelry, this group consists of amber, coral, ivory, jet, pearl, and tortoiseshell.

Origination: In this book, origination refers to the country or countries where a particular gem or mineral can be found.

Overlay: Generic term used to describe any variety of costume jewelry that involves a coating or wash over a base metal.

Oxidation: The breakdown of a metal over time as a result of exposure to oxygen and other natural elements.

Pavé: By definition, in order for a setting to be classified as pavé (pronounced "pah-vay"), one prong must touch three or more stones. If not, it is classified as a cluster. The pavé setting is most often used with diamonds and cubic zirconia.

Pavilion: The lower portion of the gemstone. It begins just below the girdle.

Pearl essence: A liquid coating that adds a pearl-like luster to simulated pearls. This coating is actually derived from the scales of a certain type of herring.

Phenomena: Special properties exhibited by certain gemstones. An excellent example of a phenomenon is the color-change property of the alexandrite.

Pleochroism: The ability of certain gems to display two or more colors when viewed from different angles.

Points: Units of measurement used to express the carat weight of a gem or simulant. One carat is equal to one hundred points, a half carat (.50) is equal to fifty points, a quarter carat (.25) is equal to twenty-five points, and so on.

Precious gem: At one time, only four gems were considered to be precious: the diamond, emerald, ruby, and sapphire. Changing conditions in the colored-gem market have pretty much made this term a thing of the past.

Precious metals: The industry defines gold, silver, platinum, and palladium as precious metals. Unlike gemstones, the term *precious* is still widely accepted when used to delineate metals.

Primary deposit: When a gem (or mineral) is found resident in its original rock.

Refraction: In gems, the bending of light as it enters the stone and slows down.

Refractive index: Developed by Willebrord Snell in the seventeenth century. This process usually incorporates a refractometer, which measures the speed and angle of light as it enters a gemstone. The refractive index is often the key to gem identification.

Rose cut: An age-old art of faceting a stone to look like an opening rose. This type of cut is usually seen in groups of six. This method of cutting is sometimes used in the creation of marcasite jewelry.

Rough: In gemology, this refers to the raw, natural state in which gems are found.

Rutiles: Needle-like inclusions (or foreign matter) within a stone, which can produce such gem phenomena as an asterism (or star) and a cat's-eye, depending on the direction of the rutiles.

Scarab: A stone carving that shows the sacred scarab beetle in intricate detail. The ancient Egyptians considered the scarab a symbol of the soul.

Secondary deposit: A deposit of gems that has been worn away from its original site, usually by the effects of weather. An alluvial deposit is one example of a secondary deposit.

Semiprecious gem: At one time, this category was reserved for all gems other than diamonds, emeralds, rubies, and sapphires. Like the category of precious gems, this term is seldom used today.

Shaving: Cutting a stone generously across the table, while allowing for very little depth. Shaving makes a stone appear to be of higher carat weight than it is.

Sheen: Another name for iridescence.

Simulant: A human-created gem having the same look, but not the same physical properties, as its natural counterpart. An excellent example is the cubic zirconia, which has a diamond-like appearance yet none of its properties.

Single cut: Stones consisting of seventeen facets or less.

Soudé: Assembled gems of multiple layers, held in place by an adhesive. A newer term sometimes used interchangeably with *triplet*.

Species: A gem with distinct characteristics that are well defined.

Specific gravity: In gemology, this refers to a process that determines the weight (or density) of a gem when compared to that of an equal volume of water. The result is a ratio expressed as a single number.

Steam cleaning: Used primarily by jewelers. This aggressive cleaning process combines steam with pressure. Steam cleaning is not suitable for all gemstones.

Step cut: A manner of cutting a gem with rectangular facets along its perimeter.

Swiss cut: Stones consisting of thirty-three facets.

Synthetic gemstone: A man-made stone that, unlike a simulant, has the same chemical composition and crystal structure as its natural counterpart.

Table: The flat top part of a gemstone.

Table facet: Central facet on the table (or crown) of a gem or simulant.

Trichroism: Those gems that display precisely three different colors when viewed from different angles.

Triplet: A man-made creation of three parts that normally includes a clear protective top layer fused together with a thinly sliced gem and a clear or colored adhesive.

Ultrasonic cleaners: Ultrasonic cleaners convert high-frequency sound waves into electrical energy, which causes a solution to bubble. This type of cleaning method is quite useful when intricate pieces of jewelry are involved. Delicate gems and organics such as amber, coral, jet, and pearls should not be cleaned in this manner.

Veins: In gemology, this usually is a term loosely used to describe long thin lines that occur on the surface of the gem. The black lines normally seen in turquoise are an excellent example of veins.

Recommended Reading Materials

They've all helped me for years. Maybe they can help you too.

Cera, Deanna Farneti. *Amazing Gems.* New York: Harry N. Abrams, Inc., 1997.

Hall, Cally. *Gemstones.* New York: Dorling Kindersley, 1994.

Holden, Martin. *The Encyclopedia of Gemstones and Minerals.* New York: Facts On File, 1991.

Kunz, George F. *The Curious Lore of Precious Stones.* New York: Dover Publications, 1970.

Lyman, Kennie, ed. *Simon & Schuster's Guide to Gems and Precious Stones.* New York: Simon & Schuster, 1986.

Matlins, Antoinette L., PG, and Antonio C. Bonanno, FGA, PG, ASA. *Jewelry and Gems: The Buying Guide.* Woodstock, Vt.: GemStone Press, 1993.

Post, Jeffrey E. *The National Gem Collection.* New York: Harry N. Abrams, Inc., 1997.

Robinson, George W., Ph.D. *Minerals.* New York: Simon & Schuster, 1994.

Schumann, Walter. *Minerals of the World.* New York: Sterling Publishing, 1997.

Sofianides, Anna S., and George E. Harlow. *Gems & Crystals from the American Museum of Natural History.* New York: Simon & Schuster, 1991.

Voynick, Stephen M. *Yogo: The Great American Sapphire.* Missoula, Mont.: Mountain Press Publishing Company, 1987.

Suggested Gemstone Periodicals

Gems & Gemology. GIA Subscription Department. 5345 Armada Drive, Carlsbad, CA 92008. Telephone: (800) 421-7250.

Jewelers Circular Keystone. PO Box 2085, Radnor, PA 19080-9585. Telephone: (609) 786-2552.

Lapidary Journal and *Colored Stone.* For both: Circulation Department, PO Box 124, Devon, PA 19333-9933. Telephone: (800) 676-4336.

In-Home Study Courses

Gemological Institute of America (GIA), 5345 Armada Drive, Carlsbad, CA 92008. Telephone: (800) 421-7250 ext. 4001; fax: (760) 603-4080; e-mail: http://www.giaonline.gia.edu.

Index

channel settings, 42, 43, 119
charoite, 135
chatoyancy (cat's-eye), 19, 43, 61, 65, 67, 70–71, 72, 88, 105, 113, 114, 118, 124, 125, 133, 169, 171, 172, 173, 177
chemicals, for enhancement, 25–26
chiastolite, 123
Chinese zodiac, 170
chromium, 39, 62, 70, 72, 78, 79, 85, 86, 92, 107, 113, 115, 126, 138
chrysoberyl, 19, 70–73, 78, 88, 98, 172; cat's-eye (color-change), 17, 19, 70–71, 105, 172, 177; yellow, 71–72, 134. *See also* alexandrite
chrysolite. *See* peridot
chrysoprase, 16, 65, 67–68, 167, 172, 174, 177
citrine, 16, 63, 65, 100, 101, 102, 110, 120, 167, 169, 172, 173, 177
clarity, of diamonds, 53
clasps, 44–45, 46, 47, 152, 163
classification of gems, 14–16
cleavage, 108, 129, 132, 179
coatings, for enhancement, 26, 111
color, 38–39, 41, 42, 53, 56–57, 137–38, 151–52
color change, 19–20, 70, 78–79, 83–84, 181
color play, 20, 93, 94, 118
color zoning, 20, 100, 115, 127
computers, settings and, 14, 24
conchiolin, 142, 143, 151, 165
copper, 19, 83, 84, 105, 116, 117, 124, 126, 129, 174
coral, 16, 23, 26, 155, 157–58, 159, 160–61, 167, 169, 171, 172, 173, 174, 177
cordierite, 128
cornelian. *See* carnelian
corundum, 27, 28, 32, 37, 38, 72–81, 88, 106, 121, 129. *See also* pad-paradscha; rubies; sapphires
crystal balls, 103
cubic zirconia, 32–33, 36, 58, 62, 93, 117, 182
Cullinan stone, 52
cut, 24, 39, 40–41, 54–55. *See also specific cuts*

danburite, 16, 27, 125–26, 174, 175, 177
dead spots, 39, 41, 55
deep-shaft mining, 22, 23
demantoid, 91, 134, 177
density, 35–36
deposits, 22–23, 50. *See also* alluvial deposits; primary deposits; secondary deposits
depth of cut, 41
desirability, of gems, 14, 15–16
diamonds, 14, 17, 36, 37, 38, 42, 43, 50–59, 60, 72, 86, 91, 93, 94, 117, 126, 141, 147, 164, 167, 169, 172, 173, 174, 175, 176, 177, 181, 182; colored, 50, 51, 56–57, 59; cut of,

40, 41, 54–55; enhanced, 28, 56–57; gem-grade, 16, 59; industrial-grade, 16, 51; paste, 58; simulants and, 58; synthetic, 57–58; white, 58–59
dichroism, 20, 64, 83–84, 124, 125
diopside, 107, 126, 134, 175, 177
dioptase, 126–27, 177
disclosure, 29–31, 153
disthene. *See* kyanite
doublets, 33–34, 94, 96
durability, 14, 15, 34, 37–38
dyes, for enhancement, 26, 28, 30

emerald cut, for diamonds, 54
emeralds, 14, 15, 16, 17, 28, 33, 39, 60, 61–63, 64, 65, 75, 89, 90, 106, 113, 126, 133, 167, 169, 172, 173, 177, 181
enhancement of gems, 24, 25–31, 56–57, 118
expensive gems, 17
extremely expensive gems, 17

facets, 14, 24, 39, 40–41, 54–55
false doublets, 33
fancy shapes, 39–40
Federal Trade Commission (FTC), 30–31, 40
feldspar, 18, 19, 20, 81–84, 109, 130, 133, 134, 174, 175, 176; oligoclase family of, 83; orthoclase family of, 18, 37, 67, 82–83; plagioclase family of, 18, 81, 83, 133. *See also* adularia; labradorite; moonstone; spectrolite; sunstone
fibrolite. *See* sillimanite
fisheye, 41
fishhook (lobster-claw) clasp, 44
fluorescence, 20, 53, 89, 125, 127
fluorite (fluorspar), 20, 37, 127, 174, 175, 177
foiling, for enhancement, 27, 58, 126
four-prong settings, 41–42
full cuts, diamonds as, 54

garnets, 16, 19, 84–92, 99, 167, 169, 173, 174, 175, 176; andradite, 91, 177, *see also* demantoid; anthill, 86; color-change, 20; grossular, 16, 78, 89–91, 177, *see also* grossularite, hessonite, rosolite, tsavorite; Malaia; 88; red, 78, 85–88, *see also* almandine, pyrope, rhodolite, star garnet; spessartine, 17, 85, 88, 172, 175, 176, 179, *see also* mandarin garnet; uvarovite, 91–92, 179
gaspéite, 135
gem grade, 16
Gemological Institute of America (GIA), 27, 45, 53
geodes, 22, 95
girdle, 41, 111

glass filling, for enhancement, 27
gold, 36, 46, 47, 145, 152, 173
Golden Giant, 52
goldenite, 105, 174, 175, 176, 177
Golden Jubilee diamond, 51 52
gold lip *(Pinctada maxima)*, 145, 146
goldstone, 19, 84, 105
goshenite, 16, 60, 63, 172, 174, 175, 177
grossularite, 89
guarantees, 47, 152, 163
gypsum, 37

hardness, 15, 36, 37, 72, 85, 90, 92, 108, 124, 125, 126, 127, 129, 130, 132, 133, 134, 137, 164, 177–79
hawksbill turtle, tortoiseshell from, 160
hawk's-eye, 19, 105. *See also* chatoyancy (cat's-eye); tiger's-eye
heart shape, 39, 145
heat, for enhancement, 27–28, 31, 56, 57, 61, 80, 102, 104, 108, 109, 111, 112, 113, 115, 118, 119, 120, 125, 131
heliodor, 16, 60, 63–64, 134, 167, 172, 174, 177
heliotrope. *See* bloodstone
hematite, 19, 66, 102, 117, 127–28, 167, 169, 172, 174, 178
herringbone chain, clasp for, 44
hessonite, 89–90, 120, 172, 175, 178
hiddenite, 107, 108, 126, 174, 175, 178
home-shopping channels, 29, 30, 31, 88, 93
Hope Diamond, 52
hornblende, 176
horsetails, 91
hyacinth, 90, 118, 119–20, 167, 169, 171, 172, 173, 178

illusion setting, 43
inclusions, 16, 19, 20, 23, 39, 54, 71, 72, 74, 77, 87, 88, 89, 91, 92, 96, 101, 105, 114, 117, 123, 133, 137, 164
indicolite, 17, 113–14, 115, 178
indigolite, 113
industrial grade, 16, 51, 134
inexpensive gems, 16
inserting-box clasp, 44
inspection of gems, 46, 55, 153
insurance, appraisals for, 45–46, 163
iolite, 16, 128, 172, 178
iridescence, 20, 65
irradiation, for enhancement, 28, 31, 109, 150, 165
ivory, 23, 155, 159–60, 161, 178

jade, 21, 26, 37–38, 43, 102, 103, 133, 136–39, 159, 167, 168, 169, 172, 173, 174, 175, 176; imperial, 138. *See also* jadeite; nephrite

Credits

All photographs courtesy of Sunil Agrawal of STS Gems–Thailand.